AF307717

REAL WORLD SPEECH PROCESSING

edited by

Jhing-Fa Wang
National Cheng Kung University
Taiwan, R.O.C.

Sadaoki Furui
Tokyo Institute of Technology
Tokyo, Japan

Biing-Hwang Juang
Georgia Institute of Technology
Atlanta, Georgia, U.S.A.

Reprinted from
a Special Issue of the
Journal of VLSI SIGNAL PROCESSING SYSTEMS
For Signal, Image, and Video Technology
Volume 36, Nos. 2-3
February-March, 2004

SPRINGER SCIENCE+BUSINESS MEDIA, LLC

Journal of VLSI
SIGNAL PROCESSING SYSTEMS
for Signal, Image, and Video Technology

Volume 36, Nos. 2/3, February/March 2004

Special Issue on Real World Speech Processing
Guest Editors: Jhing-Fa Wang, Sadaoki Furui and Biing-Hwang Juang

Library of Congress Cataloging-in-Publication Data

Real World Speech Processing / edited by Jhing-Fa Wang, Sadaoki Furui, Biing-Hwang Juang.
p.cm.
Reprinted from a special issue of the Journal of VLSI SIGNAL PROCESSING SYSTEMS for Signal, Image, and Video Technology, Volume 36, Nos. 2-3, February-March 2004.
ISBN 978-1-4419-5439-8   ISBN 978-1-4757-6363-8 (eBook)
DOI 10.1007/978-1-4757-6363-8

Permission for books published in Europe: permissions@wkap.nl
Permissions for books published in the United States of America: permissions@wkap.com

Printed on acid-free paper.

Journal of VLSI Signal Processing 36, 79, 2004

Guest Editorial

Speech is the most natural form of communication among humans. As machines become ever more capable and their use more widespread due to advances in computing, the need to allow natural communication between a human and a machine also gains critical significance. In order to realize such a system, it is essential that the speech communication process is well understood and properly modeled computationally.

Speech processing has made tremendous progress in the last 20 years of the 20th century, spurred on by advances in signal processing, algorithms, computational architectures, and hardware. Many practical applications have been demonstrated and new research directions have been identified. With the steady increase in computing power, the growing reliance on rapid electronic communication, and the maturing of speech processing technologies, we will see functional, commercially viable speech interfaces emerge, soon and ubiquitously. In the near future, speech processing is expected to flourish further and bring about an era of true human-computer interaction.

However, making speech technologies useful in the real world, particularly as a natural interface to intelligent services, is still a challenge. Compared to a human listener, for example, most of the spoken language systems do not perform well when actual operating conditions deviate from the intended ones. This special issue focuses on the most recent research and development aiming at making speech technologies applicable to the real world. The papers in this special issue cover a wide range of speech technologies including speech analysis, enhancement, recognition, and corpora for speech research. The greater part of the papers are extended version of the papers presented at International Workshop on Hands-Free Speech Communications held on April 9–11, 2001, at ATR, Kyoto, Japan. In addition, we have invited the Microsoft group to report their recent work aiming at the development of enabling technologies for speech-centric multimodal human-computer interaction.

We would like to express our sincere gratitude to the reviewers for their timely and insightful comments on the submitted manuscripts.

Jhing-Fa Wang
Sadaoki Furui
Biing-Hwang Juang

Journal of VLSI Signal Processing 36, 81–90, 2004
© 2004 Kluwer Academic Publishers.

A Robust Bimodal Speech Section Detection

KAZUMASA MURAI AND SATOSHI NAKAMURA
ATR Spoken Language Telecommunications Research Laboratories, 2-2-2 Hikaridai Seika-cho, Soraku-gun, Kyoto 619-0288, Japan

Received October 30, 2001; Revised January 2, 2003; Accepted January 29, 2003

Abstract. This paper discusses robust speech section detection by audio and video modalities. Most of today's speech recognition systems require speech section detection prior to any further analysis, and the accuracy of detected speech section s is said to affect the speech recognition accuracy. Because audio modalities are intrinsically disturbed by audio noise, we have been researching video modality speech section detection by detecting deformations in speech organ images. Video modalities are robust to audio noise, but their detection sections are longer than audio speech sections because deformations in related organs start before the speech to prepare for the articulation of the first phoneme, and also because the settling down motion lasts longer than the speech. We have verified that inaccurate detected sections caused by this excess length degrade the speech recognition rate, leading to speech recognition errors by insertions. To reduce insertion errors, and enhance the robustness of speech detection, we propose a method that takes advantage of the two types of modalities. According to our experiment, the proposed method is confirmed to reduce the insertion error rate as well as increase the recognition rate in noisy environment.

Keywords: speech detection, multimodality, face detection, motion detection

1. Introduction

It is well known that humans can take advantage of all available modalities for robust understanding. McGurk [1] showed that visual information influence speech recognition, known as the *McGurk effect*. Potamianos et al. showed that visual information aids humans in recognizing speech in noisy environments [2]. If video images are provided to a speech recognition system, and if we know how to extract the useful information out of the modalities provided, then we might be able to make the speech recognition system more robust. Our objective is to extract speech section information from audio-video modalities.

Several studies have been done on speech detection from audio modalities, and sophisticated methods have been proposed [3, 4]. However, the accuracy of detected sections continues to be intrinsically disturbed by audio noise, and experiment results in literature [3] confirm that audio noise does in fact affect detected sec-

tions. We have also confirmed that the accuracy degradation of speech detection further degrades the speech recognition rate. To enhance the robustness to audio noise, research has been carried out on bimodal speech recognition and detection [2, 5, 6]. All of these works have shown that the visual information can increase the recognition rate in noisy environments.

There are three major research areas in bimodal speech detection, i.e., face detection, feature extraction, and fusion of two modalities.

There are a number of researchers working on face detection, some of them require the skin color to be adjusted to detect the face [7], and some of them require a long computational time [8]. For our application, we have simplified a method proposed by Fukui et al. [9] to reduce the computational time. We describe the details of the method in Chapter 4.

Feature vector extraction from video modalities is also a problem of real world applications. It is obvious that humans can also observe speech by the shape

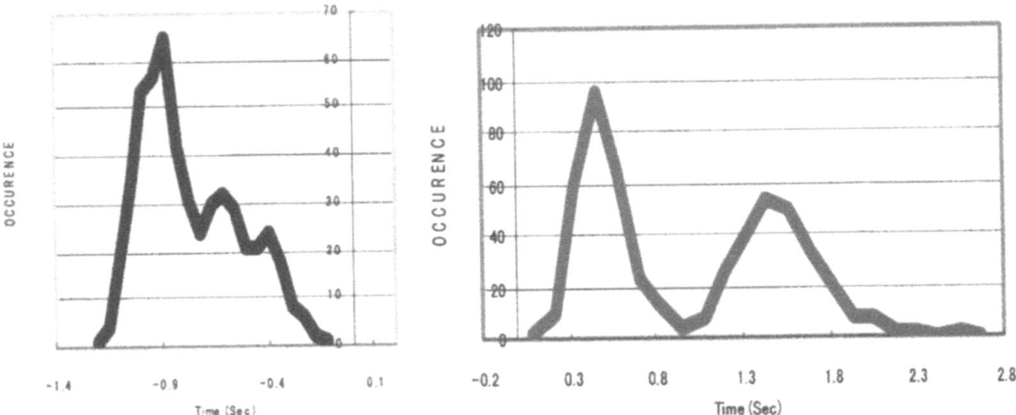

Figure 1.   Time difference between two Modalities start point (1(a), left) and end point (1(b), right).

changes of speech related organs. In fact, some methods have employed shape-based approaches by tracing the contour of the lips, or detecting specific organs. In this way, speaker independent speech recognition and detection is relatively easy because there is a straight-forward relationship between articulation (or *"viseme"*) and lip shape. However, shape tracing is a relatively expensive approach for real time application. Most of today's researches in bimodal speech recognition [2, 5] employ the Fourier transformation of two-dimensional images to extract feature vectors. These vectors have to be statistically analyzed, necessitating a larger database of feature vectors than a shape-based method. However, because our goal is to detect speech and we do not have to deal with detailed feature vectors, we simply calculate the difference between subsequent frames (described in Chapter 5) that does not require a database or shape tracing, yet retains speaker independency.

The last major problem is the fusion of two modalities. There is a difference between audio and video events in the time domain. This is because a speaker has to prepare for the articulation of each phoneme before it is made audible, and to relax the articulation after the utterance is finished. In addition, while visual modalities are influenced by preparation, articulation, and relaxation, the corresponding audio modality events might not be observable or might occur at different times. To model phoneme and viseme level speech events with differences in the time domain, Kumatani et al. proposed a speech recognition method [5] that trains visual HMMs and audio HMMs separately before composing them. Then, HMMs of these modalities are weighted with the GPD algorithm. According to our investigations, the time difference of

the start point and end point may exceed one second (Fig. 1). In this case, a sophisticated method would be too expensive to detect the speech. Instead, we simply calculate the intersection of the visual and audio speech section. Even with this simple method, we are able to enhance the word accuracy in both clean and low SNR environments. Results are shown in Chapter 6.

2.   Speech and Motion

The fact that there are some people who can communicate without audio modalities is evidence that speech is readable, and that speech and motion are related. To investigate the relationship between speech and motion, we carried out an experiment. One native Japanese speaker was instructed to read 520 isolated words separately. The speaker's voice and face images were recorded. We have carefully observed the speech and motion to identify the speech related motion. Measurable results are shown in Figs. 1 to 3.

Figure 1(a) shows the start time difference between the motion and audio modality speech. The horizontal axis shows the start time of the motion relative to the start time of the audio modality speech (time = 0 shows the start time of the audio modality speech), and vertical axis shows the occurrence. Figure 1(b) shows the end time difference between end time the two modalities (time = 0 shows the end time of the audio modality speech).

Figure 2 shows the relationship of duration between the two modalities. The horizontal axis shows the duration of the motion and the vertical axis shows the duration of the audio modality speech, which includes

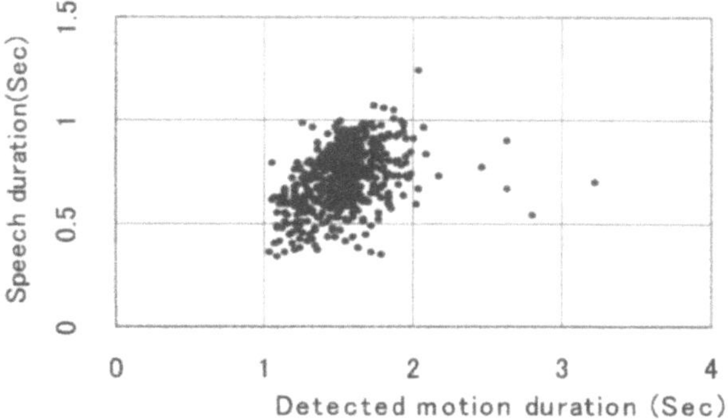

Figure 2. Duration of speech and motion. Duration of motion is longer than the speech duration. This graph shows Japanese isolated word speech detection result.

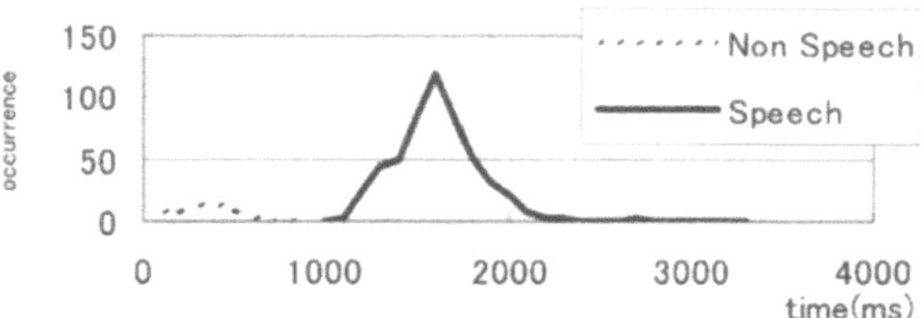

Figure 3. Duration of speech/non-speech motion.

speech of only 0.32 second. The duration of the motion with respect to the shortest speech exceeds one second.

Figure 3 shows the duration of speech and non-speech motion. The horizontal axis shows the duration, and the vertical axis shows the occurrence of the speech (solid line) or non-speech motion (dotted line). From the graph, it is clear that the durations of non-speech sections do not exceed one second. As a result of our observations, we found that:

(1) The images of lips or speech organs move as the speaker's face moves, but the shapes remain intact;

(2) The lip shape is altered not only in speech sections but also in non-speech sections (e.g., the opening/closing of the mouth for breathing), but the duration and speed are significantly less than the duration and speed during speech motion;

(3) The duration of motion is related to the duration of audio modality speech, and the speech motion always exceeds one second;

(4) Non-speech motion like the opening (closing) of the mouth for breathing does not exceed one second;

(5) Speech motion starts no later than the audio modality speech, including the words, which begin with

bilabials. And Speech motion lasts no sooner than audio modality speech, including the words, which end with the nasals.

(6) From our observations speech can be identified by the durations of shape changes.

3. Detection and Recognition

As we have mentioned in the abstract, most of today's speech recognition systems require speech section detection prior to any further analysis, and the accuracy of detected speech sections is said to affect the speech recognition accuracy. For example, if the beginning of an utterance is not detected, that part is lost before the recognition. However, it is also thought that the lengths of excess sections before or after the audio modality speech do not degrade the recognition result significantly. If this were true, the speech motion, which is longer than the audio modality speech as described in the previous chapter, would be acceptable as the speech detection result. To verify this, we carried out an experiment:

Two Japanese native speaking males were instructed to read 56 hotel reservation task utterances each, in a sound-proof room through a close-talk microphone (frontal face video was also recorded, and was used in the experiment in Chapter 6). A total of 123 utterances are recorded including incorrectly read, but meaningful ones. The audio modality SNR was 46 dB. For each utterance, the start and end points were hand labeled. To simulate detection errors at the start point, speech sections along with silence periods of 0.1, 0.2, ...2.0 second lengths prior to the hand labeled start point were cut out and recognized by speaker independent LVCSR. The detailed LVCSR is described in Chapter 6. The end point errors were also evaluated in the same manner, for 0.1, 0.2, ...2.5 second long silence periods after the hand labeled end point. The recognition results in word accuracy (solid line) and word correct rate (doted line) are shown in Fig. 4(a) (start point errors) and Fig. 4(b) (end point errors).

From the experiment results, we can say that the word accuracy significantly degrades as the simulated detection errors in both the start and end increase. The word correct rate, which does not take account of insertions, slightly degrades. As a result, the major cause of degradation is insertions, as shown by the difference between the word correct rate and word accuracy. Of special note is that, excess sections induce insertion errors, even under clean conditions. In addition, speech

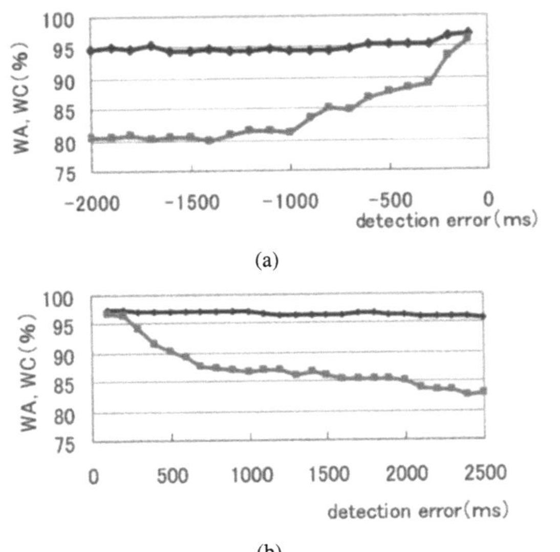

Figure 4. (a) Error induce by the speech detection (start point). (b) Error induce by the speech detection (end point).

detection by visual modalities can induce insertion errors.

To maximize the speech recognition rate, we employ two modalities.

4. Face Detection

Our speech section detection consists of the following steps:

Step 1. Detect the face by a special filter and shape;
Step 2. Detect the cross-section of the lips;
Step 3. Detect speech by the lip cross-section history.

For the source image, we use the NTSC DV format (720 by 480 pixels, 29.97 frames per second (fps), interlaced). In our interlaced video, odd lines and even lines are alternately recorded at different times. We separate the odd/even lines and interpolate each group to create a pseudo non-interlaced 59.94 (half-frame per second) video sequence before detection. The following describes each step.

4.1. Detecting the Face

There are many methods to detect the face out of a presented image [7–10]. To simplify the detection method, we assume that there is one upright face within the image. Under this assumption, the face detection is re-

duced to the case of *finding location* (x, y) *and size* (s) *of the face that maximize face likelihood function* $f(x, y, s; I)$ *within image I*, where I is a two dimensional array of R, G, B values. To detect the face in real time, and to eliminate pre-determined values for the face likelihood function, we detect the face by a special property of the facial image. Our method is based on a modified *Blob filter* [9] and a *layout mask*, which is a set of five rectangles. It uses a layout of the eyes, eyebrows, and mouth. Our implementation is as follows:

(1) Down sample the source image to 360 by 240 pixels to reduce the detection time (Fig. 5(a)).
(2) Transform color space to gray scale by *2R-G-B*, where R, G, B denote the red, green, blue values of each pixel. *2R-G-B* has higher values for Reddish and Yellowish colors (skin, lips), and near zero values for neutral colors (eyes, teeth, eyebrows, dark area), and negative values on Bluish, or Greenish colors (Fig. 5(b)). This color transformation reduces the lighting non-uniformity.
(3) Apply the modified Blob filter. Because facial organs are known to have more horizontal edges than vertical ones, we have modified the shape of the Blob filter from round to square (Fig. 6) to emphasize these edges. The filtered value $\eta(x, y)$ at the pixel location (x, y) is given by:

$$\eta(x, y) \leftarrow \frac{\sigma_b^2}{\sigma_T^2}$$
$$\sigma_b^2 \leftarrow n_1(\bar{P}_1 - \bar{P}_m)^2 + n_2(\bar{P}_2 - \bar{P}_m)^2$$
$$\sigma_T^2 \leftarrow \sum_{i=1}^{N}(p_i - \bar{P}_m)^2$$

where n_1 is the number of pixels in region 1, n_2 is the number of pixels in region 2, $N = n_1 + n_1$, and, p_i is the pixel values of the *i*th within the region 1 and region 2, $\bar{P}_1$, $\bar{P}_2$, and $\bar{P}_m$ denote the average pixel values of region 1, region 2, and (region 1 $\cup$ region 2) respectively. This filter emphasizes horizontal edges (Fig. 5(c)).
(4) Define the layout mask by a set of five rectangles $\boldsymbol{R}$, where each of the rectangle $r \in R$ is defined by location (x, y), which gives the midpoint of two corneas, and scale (s), which gives the distance between two corneas (Fig. 7). These rectangles are based on the work of Harashima et al. [11], minus the nose rectangle because the nose has less horizontal components compared to the other five areas. The face likelihood function $f(x, y, z)$ is

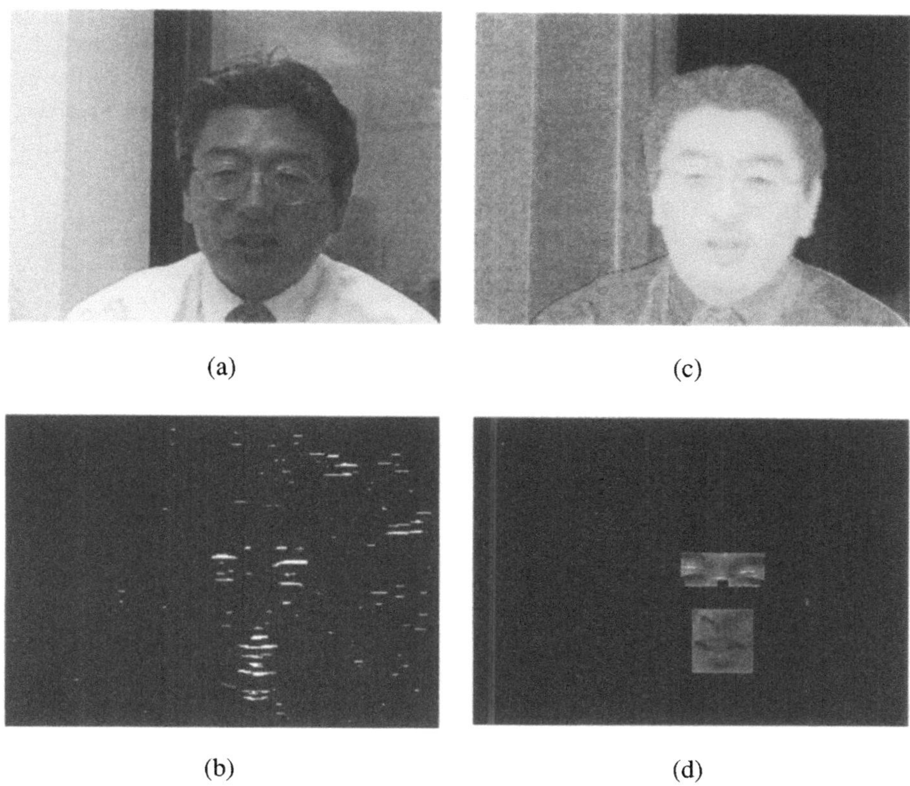

(a)   (c)

(b)   (d)

Figure 5. (a) (upper left): down sampled image, (b) (upper right): color transformed image, (c) (lower left): result of modified Blob filter and (d) (lower right): layout mask after detection.

defined by:

$$f(x, y, z) \overset{\text{def}}{\longleftarrow} \frac{1}{n} \sum_{r \in R} \left\{ \sum_{(X,Y) \in r} \eta_{(X,Y)} \right\}$$

where R is a set of five rectangles defined by (x, y, z), r is a rectangle, (X, Y) is the pixel within each rectangle r, $\eta(X, Y)$ is the filtered value at the pixel location (X, Y) of image I, and n is the total pixel count of five rectangles. By searching for the (x, y, s) giving the maximum value of

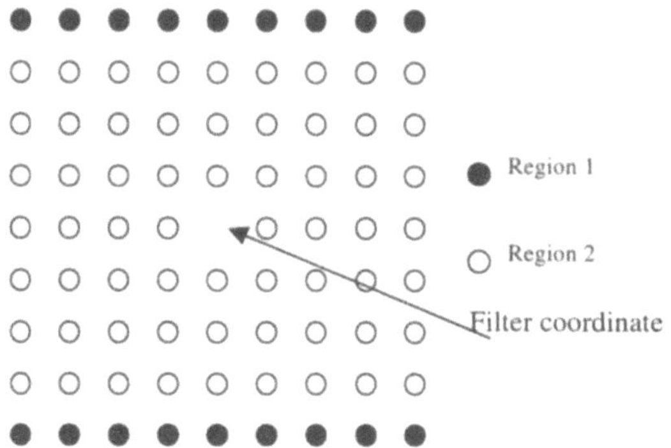

Figure 6. Modified regions applied to the Blob filter. Each circle shows the pixel location of Region 1 & 2.

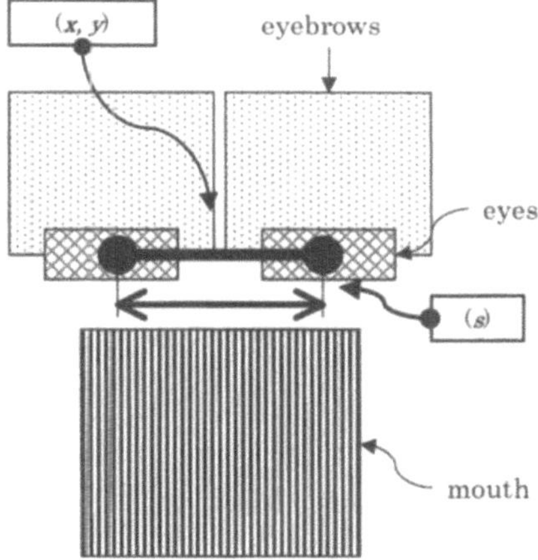

Figure 7. Layout Mask: The mask consists of 5 rectangles, 2 eyebrows, 2 eyes, and 1 mouth. Each rectangular is scaled by (s), and shifted by (z, y).

$f(x, y, s; I)$, the face can be found (Fig. 5(d)). We search for all possible (x, y, s) values for the first frame. In the following frames, we only search for neighbors $(x - 1$ to $x + 1, y - 1$ to $y + 1, s - 1$ to $s + 1)$ iteratively until the function value reaches a local maximum.

(5) Search for the corneas based on the color information. Because the eyes (corneas and white parts) comprise low saturated colors compared to their surrounding skin areas, the gravity center of each area (i.e. for each eye) is evaluated within the eye rectangle of the mask.

In this manner, the speaker's face can be found within 13 seconds, for the first frame, and within 1.3 m seconds in subsequent frames with a Pentium III 1 GHz. Except for the first frame, this searching method is applicable to real time speech detection.

4.2.   *Detecting the Cross Section of the Lips*

We assume that the speaker's face is upright, and facing the camera. In this case, the straight downward line from the mid point of the eyes crosses the lips in most cases. For further robustness, we use the perpendicular bisectors of the two corneas for the cross section line of the lips. The image along this line within the mouth rectangle is recorded. To reduce the shift of the image caused by eye blinks and detection errors, the image is shifted this cross-section image ± 25 pixels to minimize the color difference (Euclid distance of CIE L*a*b* coordinates) between subsequent frames and recorded as $Lab(f, i)$, where f is the number of frames, and i is the vertical pixel location along the line. Figure 8 shows a sample of this

record of 5.5 seconds. Because a shift of the face caused by the speaker's motion also shifts the eyes, the detected cross section of the lips is practically unaffected.

5.   Speech Detection

5.1.   *Visual Speech Detection*

As seen in Fig. 8, it seems easy to detect speech motion. To detect this, we define inter-frame energy $E(f)$:

$$E(f) \stackrel{\text{def}}{\longleftarrow} \sum_{j} \Delta\{Lab(f, j), Lab(f - 1, j)\}$$

where Δ is the color difference (Euclid distance) of the CIE L*a*b* color space. Figure 9 shows $E(f)$ of the cross-section record seen in Fig. 8. Because $Lab(f, j)$ is recorded to minimize the color difference, $E(f)$ reflects the transformation of the image. It is clear that the value of $E(f)$ has a higher value during speech. Although it seems quite easy to detect the speech from $E(f)$ with a previously determined threshold, there are also non-speech motions as described in Chapter 2 that give a higher $E(f)$ value. To identify the speech, therefore, we evaluate the duration of motion, which can clearly identify the speech out of non-speech motions as seen in Fig. 3. The experiment in Chapter 2 shows that all visually detected speech sections exceed one second for 520 Japanese isolated words (Fig. 2).

The visual detection criteria are:

- If for at least one second, $E(f)$ exceeds two times of *energy-in-non-speech*, then the first frame is detected as the beginning of detection, where

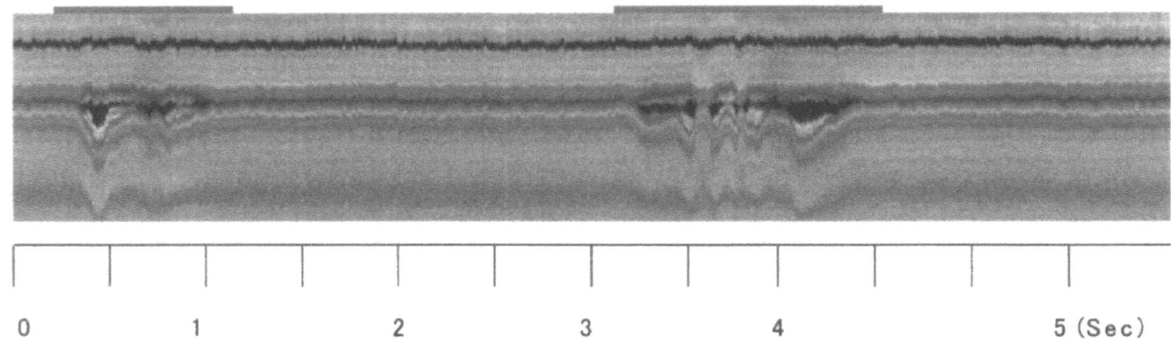

Figure 8.   The time record of lip cross section. 5.5 seconds of lip cross section $Lab(f, i)$ is recorded. The speaker spoke /haisoudesu/, and /nangmeisamadeshouka/. Upper line shows the detected speech sections.

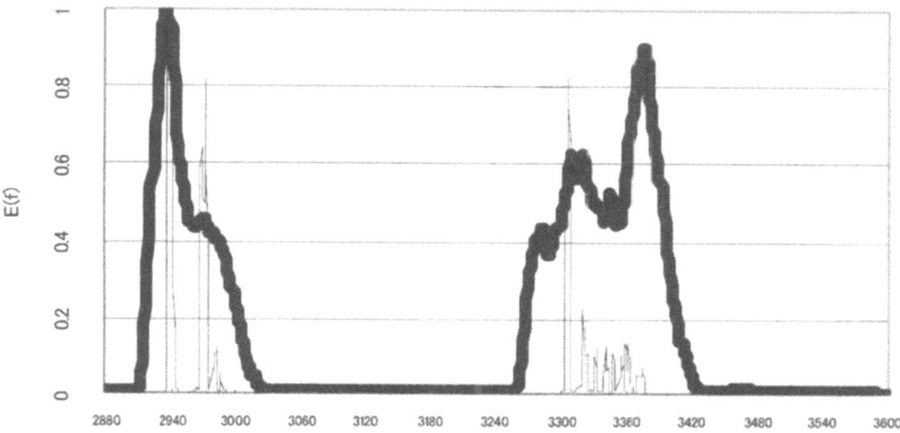

Figure 9. Interframe Energy. Bold, and thin line shows $E(f)$ with noise and audio power from the same AV source of Fig. 8. The effect of noise is invisible in this graph.

energy-in-non-speech is determined by the average energy of non-speech section for one second just before this detected section.

- If for at least 0.5 second $E(f)$ is less than two times of *energy-in-non-speech*, then the first frame is detected as the end of detection;
- An undetected section, i.e. less than 0.5 second between detected sections, is detected as a speech section.

In our detection criteria, the *energy-in-non-speech* is determined dynamically, and there is no need to previously determine the value except when the speech motion exceeds one second.

5.2. Audio Speech Detection and Fusion

As described earlier, a visually detected section is longer than an audio detected section. To reduce the excess length that may cause the insertion errors, audio modality speech detection is applied to the visually detected section. We apply a power based speech section detection algorithm ATR-EPD (ATR-EPD: speech detection module of ATR SPREC [4]). Because the objective of applying audio modality speech detection is to truncate the excess section, we simply evaluate the intersection of the speech sections in the two modalities.

6. Experiment

We evaluated the proposed method and compared it with conventional (audio or video single modality speech detection) methods.

6.1. Experimental Conditions

Two Japanese native-speaking males were each asked to sit on a fixed chair in a soundproof room, face a camera, and read 56 Japanese continuous speech utterances. The domain of the speech was a travel task. Incandescent light powered by 60 Hz AC illuminated the examinee. The speech of the examinee was recorded by an NTSC digital video camera (720×480 pixels, 29.97 fps, fixed on a stable tripod) and a close-talk microphone (48 KHz 16-bit sampling). Each pixel corresponded to a 0.5×0.5 mm area on the speaker's face. The shutter speed was adjusted to 60 Hz to eliminate flicker with the lighting. The recorded image was about 15 minutes long per examinee, and the audio SNR was 46 dB. A total of 123 utterances were recorded including incorrectly read speech. Each utterance was hand labeled. The duration of the speech based on the hand labeling ranged from 0.28 s to 10.8 s. From 10 s prior to the hand labeled start point to the 10 s after the hand labeled end point was extracted for each utterance. In addition to these files, we prepared files with audio noise adjusted to SNR 10 dB, where signal level was evaluated by the average of each hand labeled speech section, and the level of noise is adjusted by each utterance. The noise was exhibition hall noise [12].

The detected speech sections were recognized by HTK 3.0. For an acoustic modeling, a 26-phoneme, 25-dimensional (12-dimensional mel-cepstral coefficients, 12-dimensional first-order derivatives of mel-cepstral coefficients, and 1-dimensional first-order derivative of logarithmic power) value was computed with a 20 ms window length and 10 ms frame shift. CMN was applied to both the test set and acoustic model. The acoustic model comprised shared-state HMMs (1,400 states in total) with five Gausian mixture components per state. These HMMs were trained using 16 kHz 16-bit sampling clean audio data of 167 male speakers. The word dictionary consisted of 32,304 Japanese words, and the language model was a bigram model trained by the travel task.

6.2. *Evaluation Results*

We evaluated the speech detection methods by the word accuracy (Table 1), word correct rate (Table 2), and insertion count (Table 3), where the word correct rate is defined by:

Word correct: $c / (c + s + d)$
Word accuracy: $(c - i) / (c + s + d)$

where c is the number of correctly recognized words, s is the number of substituted words, d is the number of deleted words and i is the number of inserted words. The proposed method and conventional (audio or video single modality) methods were applied to the test data.

From the results, the proposed method outperformed single modality methods in all cases except the word correct rate.

Table 1.   Word accuracy (%) of each method.

Word acc%	Proposed	Audio	Video
Clean	92.3	90.4	86.2
SNR 10 dB	70.2	66.8	52.6

Table 2.   Word correct rate (%) of each method.

Word cor%	Proposed	Audio	Video
Clean	95.5	95.8	93.7
SNR 10 dB	80.2	80.5	75.3

Table 3.   Insertion (number of words per 123 utterance).

Insertions	Proposed	Audio	Video
Clean	41	71	91
SNR 10 dB	126	190	273

7. Discussions

Because audio modalities are fragile to audio noise, we additionally introduced video modalities into speech detection. However, to take advantage of the additional modalities, we have to handle huge amounts of data and face and motion recognition. This is because video modalities are relatively expensive to handle (e.g., the amount of data per second is almost 1,000 times more than the audio modalitiy alone). The experiment results shown in Chapter 3 and 6 show that the speech detection accuracy affects the overall speech recognition performance, and we confirm that video modalities improve the recognition rate, and are worthy of inclusion. We have carefully chosen the Blob filter shape and the mask shape for face detection; we can employ visual modalities and bimodal speech detection for real-time speech processing.

In our experiment in Chapter 3, the lengths of excess sections before or after audio modality speech induce insertion errors even under clean conditions. One of the reasons might be the mismatch induced by the acoustic model specially for "silence". Considering this, we reviewed some of the audio data used to train the acoustic model. The training set included audio data with an SNR of around 33 dB, while the test set included data with an SNR of 46 dB. Both audio data were clean, but signal level of silence differed by 13 dB, possibly causing the mismatch leading to the insertion errors. However, it is not always possible to determine beforehand the noise level of the environment that a speech recognition system will be used in, thereby causing the mismatch of the acoustic model. Even in this case however, the proposed speech detection can improve the recognition rate (the experiment which used an SNR of 10 dB for the audio modality test set featured training by data with an SNR of around 33 dB).

We analyzed cause of the insertions. The audio modality speech detection occasionally detected the non-speech sections as speech sections under the SNR 10 dB environment. The detected noise-only sections sometimes generated the recognition results (i.e., insertion errors). Then, by the proposed method, these non-speech sections were inhibited by the video modality detection, yielding less insertion errors. Even by only computing the intersection of the two types of modality detections, it was possible to reduce insertion errors. We are planning to improve the fusion of the two types of modalities in a more efficient way. While the audio modality speech detection requires the adjustment of some variables for clean or SNR 10 dB detection, our implementation of the visual modality speech detection does not require the adjustment of any variables. It might be possible to automatically adjust the variables of the audio modality speech detection by visually detected speech sections or vice versa.

Our implementation of the video modality speech detection gives very high accuracy in the identification of non-speech sections, i.e., we did not observe any speech (not only read style utterances, but also spontaneous conversations between examiner and examinee found in between read style utterances) when the visual modality speech detection identifies sections as non-speech sections. It might also be possible to adapt the acoustic model to the background noise during visually identified non-speech sections to further improve the recognition rate.

8. Conclusions

We proposed a robust bimodal speech section detection method. An experiment showed speech motion sections to be longer than audio modality speech sections. Another experiment showed that an excess length induces insertion errors. To detect video modality speech sections, we have proposed a method to detect the face, and a method to detect the speech from the lip cross-section image history. To reduce errors, we proposed a method that uses audio modality speech detection along with video modality speech detection. According to evaluation experiments, the proposed method reduces insertion errors compared to audio or video single modality speech detection methods in clean and SNR 10 dB environments.

Acknowledgments

The authors thank Dr. Seiichi Yamamoto, president of ATR Spoken Language Translation Laboratories, for giving us the opportunity to carry out this research.

References

1. D.G. Stork and M.E. Hennecke, *Speechreading by Humans and Machines*, Berlin, Springer, 1996.
2. G. Potamianos, C. Neti, G. Iyengar, and E. Helmuth, "Large-Vocabulary Audio-Visual Speech Recognition by Machines and Humans," in *Proc. Eurospeech 2001*, 2001, pp. 1027–1030.
3. J.C. Junqua, B. Reaves, and B. Mak, "A Study of Endpoint Detection Algorithms in Adverse Conditions: Incidence on a DTW and HMM Recognizer," in *Proc. Eurospeech 1991*, 1991, pp. 1371–1374.
4. M. Naito, H. Singer, H. Yamamoto, H. Nakajima, T. Matsui, H. Tsukada, A. Nakamura, and Y. Sagisaka, "Evaluation of ATRSPREC for Travel Arrangement Task," in *Proc. ASJ Fall Fall Meeting 1999*, 1999, pp. 113–114.
5. K. Kumatani, S. Nakamura, and K. Shikano, "An Adaptive Integration Method Based on Product HMM for Bi-Modal Speech Recognition," in *Proc. HSC2001*, 2001, pp. 195–198.
6. K. Murai, K. Kumatani, and S. Nakamura, "A Robust End Point Detection by Speaker's Facial Motion," in *Proc. HSC2001*, 2001, pp. 199–202.
7. W.H. Lin and K. Sengupta, "Manipulation of Remote 3D Remote Avatar Through Facial Feature Detection and Real Time Tracking," in *Proc. ICME 2001*, 2001.
8. H.A. Rowley, S. Baluja, and T. Kanade, "Neural Network-Based Face Detection," *IEEE PAMI*, vol. 20, no. 1, 1998, pp. 23–38.
9. K. Fukui and O. Yamaguchi, "Facial Feature Point Extraction Method Based on Combination of Shape Extraction and Pattern Matching," *IEICE D-II*, vol. J80-DII, no. 8, 1997, pp. 2170–2177.
10. http://www.cs.rug.nl/~peterkr/FACE/frhp.html.
11. Harashima et al., "Facial Image Processing System for Human-like "Kansei" Agent," *IPA*, 1998.
12. Noise Database Published by Japan Electronics and Information Technology Institutes Association, 1990.

Kazumasa Murai received M.Eng. degree from Gunma University in 1986. He worked with Cooperate Research Laboratories, Fuji Xerox 1989–2000, with ATR Spoken Language Translation Laboratories 2000–2002, with IT Media Laboratory from 2002. He received best paper award from Imaging Society of Japan. His Research interests include facial image processing and multi-modal speech detection.
kazumasa.murai@slt.atr.co.jp

Satoshi Nakamura received the B.S. degree in electronics engineering from Kyoto Institute of Technology in 1981 and the Ph.D.

degree in information science from Kyoto University in 1992. Between 1981–1986 and 1990–1993, he worked with the Central Research Laboratory, Sharp Corporation, Nara, Japan, where he was engaged in speech recognition research. During 1986–1989, he was a researcher of the speech processing department at ATR Interpreting Telephony Research Laboratories. From 1994–2000, he was an associate professor of the graduate school of information science, Nara Institute of Science and Technology, Japan. In 1996, he was a visiting research professor of the CAIP center of Rutgers, the state university of New Jersey, USA. He is currently the head of Department 1 in ATR Spoken Language Translation Laboratories, Japan. He also serves as a guest professor for Toyohashi University of Technology and Ritsumeikan University from April 2002. His current research interests include speech recognition, speech translation, spoken dialogue systems, stochastic modeling of speech, and microphone arrays. He received the Awaya Award from the Acoustical Society of Japan in 1992, and the Interaction2001 best paper award from the Information Processing Society of Japan in 2001. He is a member of the Acoustical Society of Japan, Institute of Electrical and Electronics Engineers (IEICE), Information Processing Society of Japan, and IEEE. He is currently a member of the Speech Technical Committee of the IEEE Signal Processing Society and an editor for the Journal of the IEICE Information and System Society.

nakamura@slt.atr.co.jp

Journal of VLSI Signal Processing 36, 91–104, 2004

Acoustic Feature Analysis and Discriminative Modeling of Filled Pauses for Spontaneous Speech Recognition

CHUNG-HSIEN WU AND GWO-LANG YAN

*Department of Computer Science and Information Engineering, National Cheng Kung University,
Tainan, Taiwan, Republic of China*

Received October 30, 2001; Revised June 5, 2002; Accepted September 3, 2002

Abstract.   Most automatic speech recognizers (ASRs) concentrate on read speech, which is different from spontaneous speech with disfluencies. ASRs cannot deal with speech with a high rate of disfluencies such as filled pauses, repetitions, lengthening, repairs, false starts and silence pauses. In this paper, we focus on the feature analysis and modeling of the filled pauses "ah," "ung," "um," "em," and "hen" in spontaneous speech. Karhunen-Loéve transform (KLT) and linear discriminant analysis (LDA) were adopted to select discriminant features for filled pause detection. In order to suitably determine the number of discriminant features, Bartlett hypothesis testing was adopted. Twenty-six features were selected using Bartlett hypothesis testing. Gaussian mixture models (GMMs), trained with a gradient decent algorithm, were used to improve the filled pause detection performance. The experimental results show that the filled pause detection rates using KLT and LDA were 84.4% and 86.8%, respectively. A significant improvement was obtained in the filled pause detection rate using the discriminative GMM with KLT and LDA. In addition, the LDA features outperformed the KLT features in the detection of filled pauses.

Keywords: filled pause, disfluency, Guassian mixture model, speech recognition, Karhunen-Loéve transform, linear discriminant analysis

1. Introduction

The recent growing demand for automatic speech recognizers (ASRs) has been manifested in applications such as dialog systems, call managers and weather forecasting systems. The most noticeable problem for these systems is the poor recognition rate for filled pauses because spontaneous speech is punctuated with and interrupted by a wide variety of seemingly meaningless words. This disfluent speech contains filled pauses, repetitions, lengthening, repairs, false starts, and silence pauses. All kinds of disfluencies generally destroy the smooth speaking style of speech and therefore degrade the performance of ASR. Past work on the detection of disfluent speech was based on acoustic, language, and prosodic information models. In the acoustic model approach, most of the research treated the disfluency as a general resembling recogni-

tion model [1, 2]. However, these speech recognizers are typical HMM based and accept only fluently read or planned speech without disfluencies. These works suffer difficulties in dealing with filled pauses and word lengthening because the duration of a phone tends to lengthen differently compared to general speech. For the language model-based approach, some previous research [3–5] tried to correct the recognition errors caused by disfluency using language models. These works either took the difluency into account or skipped the disfluency in the language model. This is not effective for dealing with filled pauses because they can be inserted at arbitrary positions. The corpus-based language models cannot completely model all of the various kinds of disfluent conditions. The skip strategy cannot preserve the spontaneous speech structure from destruction because it ignores the important roles of disfluent words [4, 6, 7]. Other research [8, 9] analyzed the prosody of disfluent speech. Prosodic cues such as pitch and duration were exploited to derive rules and features to detect the disfluent positions in speech. These rules and features using only pitch and duration are still not enough to model all of the types of filled pauses.

In this paper, the filled pauses "ah," "ung," "um," "em," and "hem" are investigated. The approach used for detecting these pauses is based upon two principles. The first principle was to prevent ASR performance degradation due to the presence of filled pauses. The second important principle was that filled pauses play valuable roles, such as markers of discourse structure [7], thinking and helping a speaker maintain his turn in the conversation and meaningful oral communication. Filled pauses can be characterized by two acoustic properties according to our investigation. These properties are the nasal effect and lengthening. Many salient features for describing these phenomena have been analyzed and proposed in [10–14]. For example, nasal sounds are characterized by the first two formant frequencies (at about 300 and 1000 Hz). The mean magnitude difference between the amplitudes of formant 1 and antiformant 1 is greater for the oral vowel than the nasal vowel [10, 11]. The distinctive spectral traits of nasal consonants are low first formant with higher intensity than the upper formants and a set of weak formants at 300–4000 Hz [12, 13]. The magnitude ratio of formant 2 to formant 1 is smaller in the nasal consonant [14]. Lengthening is characterized by a steady spectrogram. The spectral/cepstral coefficients used to model the vocal tract are chosen as

the features for detection. In our approach, 48 features that convey the properties of the filled pauses described above are extracted first. The Karhunen-Loéve transform (KLT) and linear discriminant analysis (LDA) are then performed on these features. The Bartlett hypothesis testing [15] was adopted to determine the number of discriminant features. Twenty-six features were selected using the Bartlett hypothesis testing under a confidence level 97.5%. Using these selected features, the filled pause detection rate was about 76.8% for the KLT features and 78.1% for the LDA features. To increase the discriminability, these features were modeled further using Gaussian mixture models (GMMs) [16], whose weights are estimated using a gradient decent training algorithm that recursively minimizes the detection error rate. An optimal threshold is set to achieve the best detection rate.

The proposed architecture for filled pause detection is shown in Fig. 1. In the training process, the speech features are extracted and then analyzed using the Karhunen-Loéve transform and linear discriminant analysis. The feature number was determined using the Bartlett hypothesis testing. These features were then modeled using the filled pause GMMs to form a model database. Similarly, the same features for the normal or fluent speech were modeled using a fluency GMM. To increase the discriminability, the GMMs were trained using a discriminative algorithm to optimize the detection rate. In the filled pause detection process, a threshold is then defined to detect the filled pauses by the verification scores calculated from the GMMs.

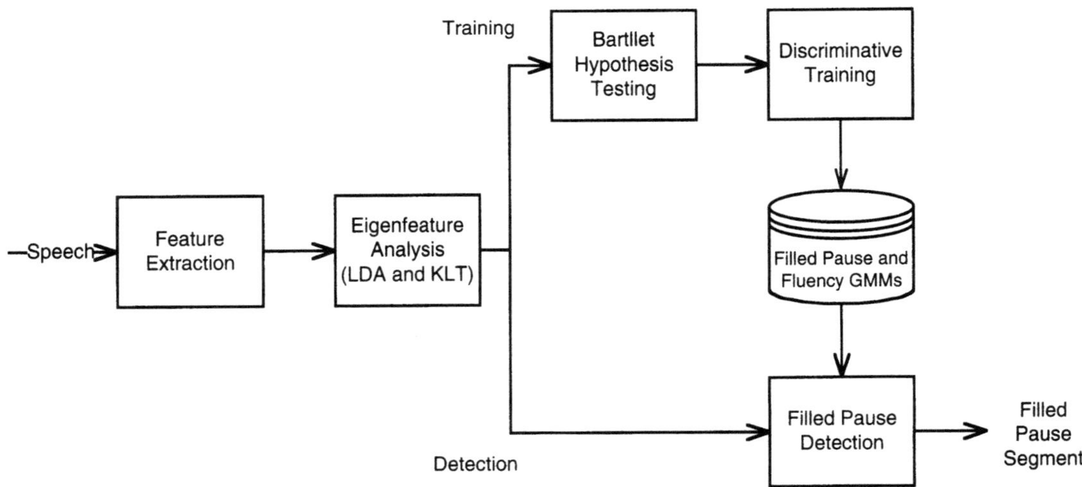

Figure 1. System architecture for the detection of filled pauses.

This paper is organized as follows. In Section 2, we describe the features chosen for characterizing the filled pauses. The Karhunen-Loéve transform and linear discriminant analysis methods are also described. In Section 3, the architecture and the GMM discriminative training are depicted. The experimental results are shown in Section 4. A brief conclusion is presented in Section 5.

2. Feature Analysis of Filled Pauses

Because filled pauses may appear anywhere in an utterance when people are talking with each other, a database of spontaneous conversations was collected from several spontaneous conversations. The spontaneous speech database contains over 8 hours of recorded speech, spoken by over 40 speakers of both sexes. These spontaneous speech utterances were transcribed, segmented and tagged into 2,160 sentences. According to our preliminary observation from the database, the filled pauses can be summarized as having two properties: the nasal effect and lengthening properties. The feature analysis and selections for these two properties are described in the following sections.

2.1. Lengthening Property

All filled pauses in spontaneous speech have a common property: lengthening. For voice lengthening, the vocal cord vibrates periodically and the vocal tract is maintained in a relatively stable configuration throughout the utterance. In other words, the produced voice changes smoothly. Figures 2(a) and (b) show the waveform and spectrogram of the speech utterance "嗯...你好" (um ... how are you) in Mandarin. The lengthening voice "um" occurs at the beginning of the utterance. The spectrogram is nearly steady compared to the voice in the middle of the utterance. Two kinds of spectral/cepstral coefficients are employed to characterize this property. The first kind of coefficient, which is famous in modeling the vocal tract, is the linear predictive coding (LPC) coefficient. The second feature, which considers the human hearing perception and has been proven useful for speech recognition, is the mel-frequency cesptrum coefficient (MFCC). We choose 12 MFCCs, 12 delta MFCCs, and 12 LPCs as the features used to detect and analyze the lengthening property. Figure 2(c) shows an example of these features, in which 12 MFCCs and 12 delta MFCCs are stable in the lengthening part.

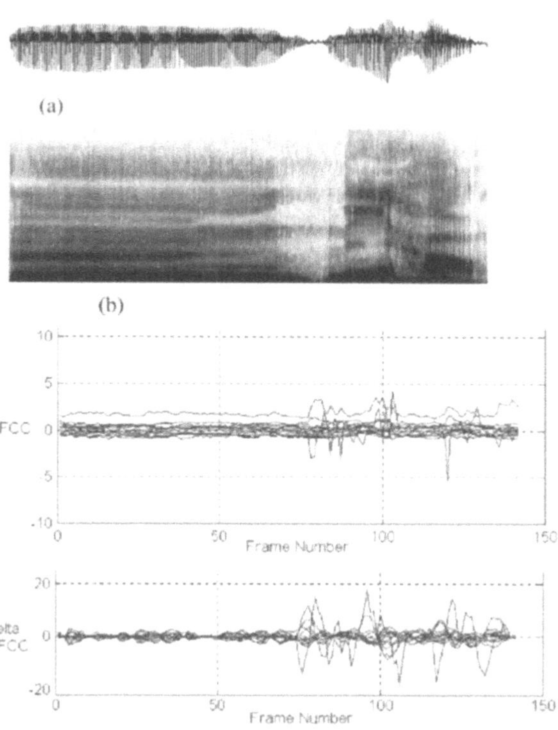

Figure 2. (a) Waveform, (b) spectrogram and (c) 12 MFCCs and 12 delta MFCCs of the utterance "嗯...你好" (um ... How are you).

2.2. Nasal Effect Property

The second property in filled pauses is the nasal effect. This property is found in the filled pauses "ah," "ung," "um," "em," and "hem." In the production of nasal sounds, the resonance characteristics are conditioned by the oral cavity characteristics forward and backward from the velum and by the nasal tract characteristics from the velum to the nostrils. A special production procedure causes this particular formant change. Some research [10, 11] has reported on nasalized voices. The noticeable cues are the first two formants (at about 300 and 1000 Hz). The first two formant frequencies for normal sounds are generally at about 250–800 and 700–2500 Hz. Figure 3 shows an example of "a" and its nasalized sound "ah." We can see that the formant frequencies of F1 and F2 for /ah/ are changed to about 250–800 and 700–2500 Hz compared to the oral sound /a/. Figure 4 shows the distributions of F1 and F2 for the vowels "a," "i," "u," "e," and "o," and the filled pauses "ah," "ung," "um," "em," and "hem." The "x" marks, representing the filled pauses "ah," "ung," "um," "em,"

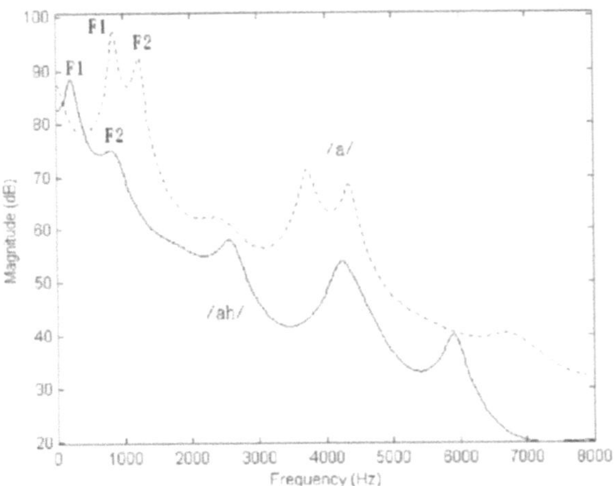

Figure 3. Spectral envelopes of /a/ (dash) and /ah/ (solid).

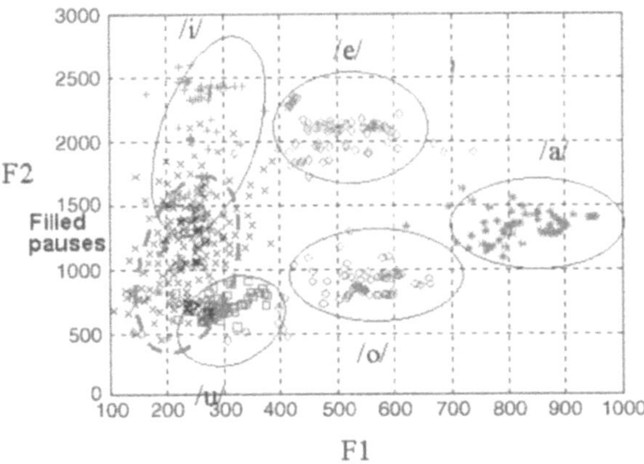

Figure 4. Plot of F2 versus F1 for vowels "a," "i," "u," "e," and "o," and filled pauses "ah," "ung," "um," "em," and "hem."

and "hem" in this figure, can be distinguished from the vowels in the vowel triangle [10]. The formant 1 (F1) and formant 2 (F2) frequencies for "ah," "ung," "um," "em," and "hem" are useful for characterizing the nasal sounds. They were also chosen as the features for filled pause detection. We also include formant 3 (F3) for analysis. In this paper, ARMA framework with modified recursive least square algorithm [17] is applied to explore the spectral information of a time varying signal and track adaptively and directly the polynomial zeros and poles of the transfer function. The average frequencies of F1, F2, and F3 for "ah," "ung," "um," "em," and "hem" are listed in Table 1 calculated from our corpus described in Section 5.1.

The second cue for nasal sounds is the mean magnitude difference between the amplitudes of F1 and

Table 1. The formant frequencies F1 and F2 of "ah," "ung," "um," "em" and "hem".

	ah	ung	um	em	hem
Formant 1	233 Hz	239 Hz	245 Hz	191 Hz	203 Hz
Formant 2	1268 Hz	1021 Hz	1253 Hz	720 Hz	778 Hz

Table 2. Average mean magnitude differences between F1 and Z1 for (a) /a/ compared to their nasalized counterparts /ah/ and (b) /e/ compared to its nasalized counterpart /em/.

	(a)		(b)	
	/a/	/ah/	/e/	/em/
Mean magnitude difference (db)	17.11	8.87	14.74	9.56

antiformant 1 (Z1) frequencies. If we denote the spectrum envelope function as $SE(f)$ where f is the input frequency, the mean magnitude difference is defined as

$$MD(F1, Z1) = SE(F1) - SE(Z1) \qquad (1)$$

In Table 2, the average mean magnitude differences for /a/ compared to its nasalized counterpart /ah/ and /e/ compared to its nasalized counterpart /em/ are estimated and compared. It is clear that the mean magnitude difference for oral vowel is greater than that for nasal vowel [10, 11]. To include more features for further analysis, $MD(F1, Z1)$, $MD(F2, Z2)$, and $MD(F3, Z3)$ are also taken into account.

The third cue for the nasal effect in filled pauses generally occurs in nasal consonants such as "ung" and "um" [11–13]. The characteristic of the nasal consonant is its distinctive spectral traits. According to the reports, the distinctive spectral traits of nasal consonants are (1) low first formant with higher intensity than the upper formants and (2) low amplitudes for the upper formants. The properties described above can be observed in two parts. The first is the formant magnitude ratio, which is defined in the following.

$$FMR(F2, F1) = \frac{SE(F2)}{SE(F1)}, \qquad (2)$$

where $SE(F1)$ represents the magnitude of F1. The equation formulates the degree of the decrease in magnitude when the voice is produced through the nostrils. Figure 5 shows a histogram of the formant magnitude ratio for "ung" and "um." The mean of this distribution is about 0.08. Figure 6 shows a histogram of the

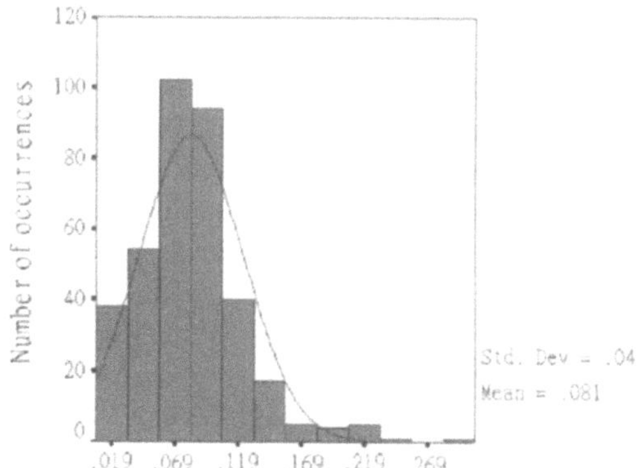

Figure 5. Histogram of formant magnitude ratio for "ung" and "um".

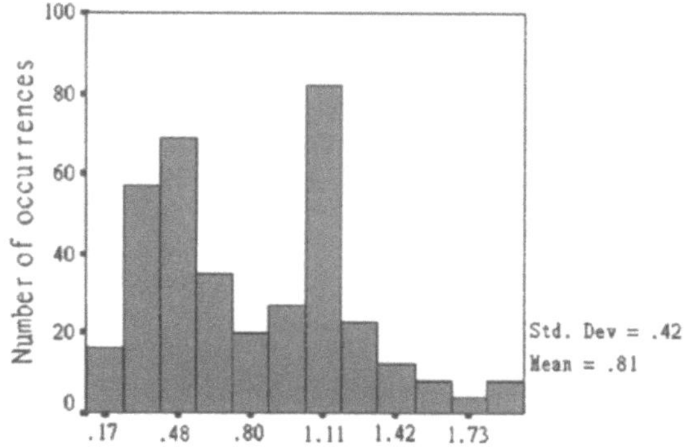

Figure 6. Histogram of formant magnitude ratio for normal voices.

formant magnitude ratio for oral voices. The distribution mean is about 0.8. The formant magnitude ratio is very special for the nasal characteristics of "ung" and "um" and can be used to characterize the property "low first formant with higher intensity than the upper formants." Similarly, the features $FMR(F2, F1)$, $FMR(F3, F1)$ and $FMR(F2, F3)$ were also included for analysis.

The fourth cue for the nasal effect is the low amplitudes for the upper formants. This also describes the dramatic decrease in magnitude when the voice is produced through the nostrils. Therefore, $SE(F1)$, $SE(F2)$ and $SE(F3)$ were also chosen as features for analysis.

3. Discriminant Feature Analysis and Selection

According to our previous analysis, there are 48 features that are possibly useful for filled pause detection. They are 12 MFCCs, 12 delta MFCCs, 12 LPCs, F1, F2, F3, $MD(F1, Z1)$, $MD(F2, Z2)$, $MD(F3, Z3)$, FMR (F2, F1), FMR (F3, F1), FMR (F3, F2), $SE(F1)$, $SE(F2)$, and $SE(F3)$. In practice, however, this number of features is too large to allow robust and fast detection. It is desirable to obtain a suitable set of features to achieve an acceptable performance. A common way to resolve this problem is to use dimensionality reduction techniques and discriminant analysis. Two of the most popular techniques for this purpose are: Karhunen-Loéve transform and linear discriminant analysis. In the following sections, we will describe how these two techniques were applied to the filled pause detection problem and how to determine a suitable feature number m to achieve an acceptable result.

3.1. Karhunen-Loéve Transform

Efficient salient feature selection is an important issue in classification. The Karhunen-Loéve transform (KLT), also known as the principal component analysis or "eigenfeatures," was adopted to choose the most efficient features for face recognition and image retrieval [18, 19]. In this approach, KLT is employed to remove information redundancies and generate orthogonal features so that the selected features are less confusing and more discriminant. Given an n-dimensional vector X, it can be expanded and approximated by

$$X = VY \qquad (3)$$

where the columns of the $n \times n$ square matrix V are an orthogonal basis vector with $V^T V = I$; Y is a transformed feature vector of the random vector X. In order to derive the efficient features for discrimination, the approximation of X using $m < n$ columns of V gives

$$\hat{X}(m) = \sum_{i=1}^{m} y_i v_i \qquad (4)$$

where v_i $(i = 1..m)$ are the column vectors of V. In other words, this is the projection of X onto the subspace spanned by the m orthonormal eigenvectors. The approximation estimation can be defined according to the mean-square error as

$$Q(X, \hat{X}) = E[\|X - \hat{X}(m)\|^2]. \qquad (5)$$

Our goal now is to choose eigenvectors that minimize the mean-square error Q. From Eq. (5) and taking into account the orthonormality property of the eigenvectors, we have

$$E[\|X - \hat{X}(m)\|^2] = \left[\left\|\sum_{i=m+1}^{n} y_i a_i\right\|^2\right]$$

$$= E\left[\sum_i \sum_j (y_i v_i^T)(y_j v_j)\right]$$

$$= \sum_{i=m+1}^{n} v_i^T E[XX^T] v_i \qquad (6)$$

Combining Eq. (6) and the eigenvector definition, we finally get

$$E[\|X - \hat{X}(m)\|^2] = \sum_{i=m+1}^{n} v_i^T \lambda_i v_i$$

$$= \sum_{i=m+1}^{n} \lambda_i \qquad (7)$$

According to the result from Eq. (7), the mean-square error is proportional to the summation of the smallest eigenvalues from $m + 1$ to n. Consequently, the chosen $v_1, v_2, .., v_m$ eigenvectors are associated with the m largest eigenvalues in the covariance matrix of X, defined as

$$\sum_X = E[(X - \mu_X)(X - \mu_X)^T], \qquad (8)$$

where μ_X is the mean vector of X. The transformed features $y_1, y_2, \ldots, y_m$ are then computed as follows.

$$y_i = v_i^T(X - \mu_X), \quad i = 1, 2, .., m. \qquad (9)$$

These features are the optimal features for producing the minimum mean-square error. We call these features the KLT features, employed for the acoustic discriminative modeling of filled pauses in the following experiments. In this approach, several sets of discriminant features such as MFCCs, LPCs, formant frequencies are not in a same magnitude, so we adopt correlation matrix instead of covariance matrix defined in Eq. (8) before applying the KLT or LDA. The correlation matrix is defined as:

$$R_X = E\left[\left(\frac{X - \mu_X}{\sigma_X}\right)\left(\frac{X - \mu_X}{\sigma_X}\right)^T\right] \qquad (10)$$

where σ_X is the standard deviation vector of X.

3.2. Linear Discriminative Analysis

Linear discriminant analysis (LDA) is another famous feature selection method [19]. It has been widely used to search for those vectors in the underlying space that best discriminate among classes. In other words, LDA creates a linear transform of these features to maximize the largest mean differences between the desired classes. The following matrices for the random vector X are defined as

1. Within-class scatter matrix:

$$S_w = \sum_{i=1}^{C} P_i E\left[(X_i - \mu_{X_i})(X_i - \mu_{X_i})^T\right] \qquad (11)$$

where X_i is the i-th class of $X(i = 1..C)$, P_i is the a priori probability of class i and $P_i \cong n_i/N$ is computed by the number n_i of samples belonging to class i divided by the total number N.

2. Between-class scatter matrix:

$$S_b = \sum_{i=1}^{C} P_i(\mu_{X_i} - \mu_X)(\mu_{X_i} - \mu_X)^T \qquad (12)$$

where μ_{X_i} is the sample mean of class i and μ_X is the global mean vector and accumulated using

$$\mu_X = \sum_{i=1}^{N} P_i \mu_{X_i} \qquad (13)$$

The goal is to maximize the between-class measure while minimizing the within-class measure. A number of different criteria were defined using various combinations of these scatter matrices in a "trace" or "determinant" formulation. One way to do this is to maximize the ratio $\frac{\det(S_b)}{\det(S_w)}$ [19]. This ratio has proven that if S_w is a nonsingular matrix, then this ratio is maximized when the column vectors of the projection matrix are the eigenvectors of $S_w^{-1} S_b$ associated with the largest eigenvalues. We call these features the LDA features. Take the filled pause "ung" as an example. In Fig. 7, the projection similarities are introduced to explain the discriminant property between "ung" and not-"ung." The horizontal axis is the projection similarity to the sound "ung" and the vertical axis is the projection similarity to the sound not-"ung." Figure 7(a) shows a distribution of the original features and Figs. 7(b) and (c) show the distributions of the features using KLT and LDA ($m = 26$), respectively. From this figure, the KLT and LDA features give better separability than the original

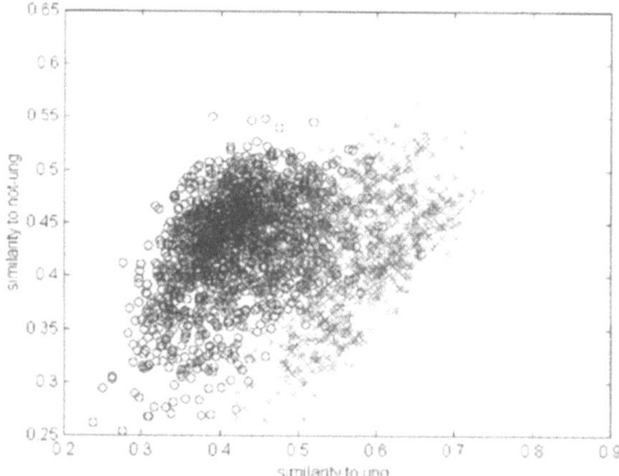

(a)

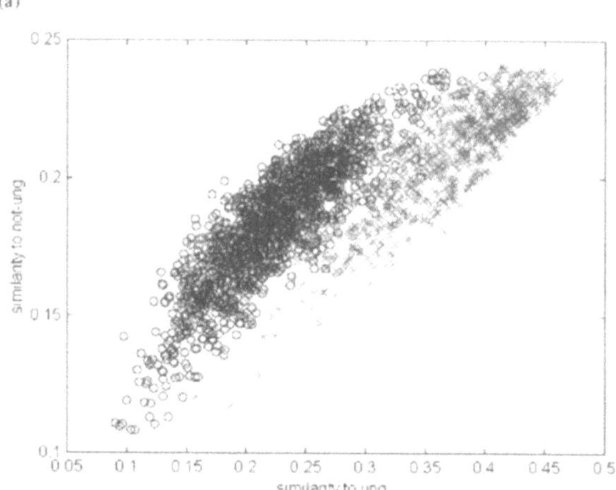

(b)

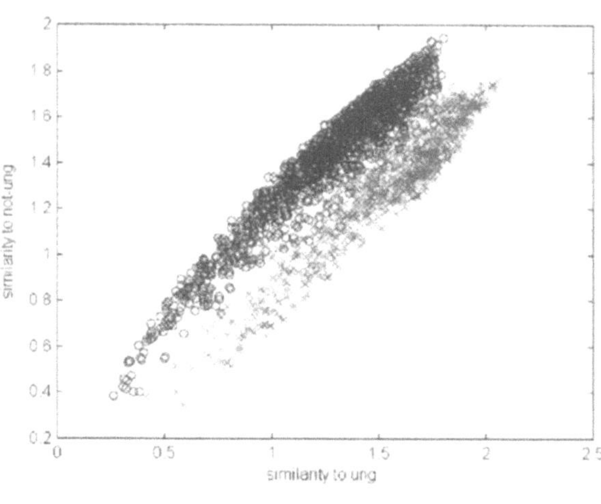

(c)

Figure 7. Comparison of the discriminability for "ung" and not-"ung" using (a) original features, (b) KLT features, and (c) LDA features. Circle: Filled pause. Cross: not-filled pause.

features when a suitable feature number is selected. It is because the KLT and LDA projections are likely to achieve better separability for a selected feature number. In our approach, for comparison between the KLT and LDA features, the values of m were chosen as the same.

3.3. Bartlett Chi-Square Testing

Determining the feature number m is another important task. In this section, our motivation is based on selecting a suitable m with an acceptable performance without paying too much effort to run all the experiments for m from 1 to 48. The following two methods, conventional and statistical methods, are adopted and compared. The conventional method utilizes the residual error to evaluate the efficiency for different numbers of features. If we rank the eigenvalues of correlation matrix R_X to $\lambda_1, \lambda_2, \ldots \lambda_n$, according to Eq. (7), the residual error is simply defined according to the sum of the $n - m$ smallest eigenvalues not used, compared to the total sum of eigenvalues. It is written as

$$\text{Residual Error} = \frac{\sum_{i=m+1}^{n} \lambda_i}{\sum_{i=1}^{n} \lambda_i} \quad (14)$$

The residual error is always chosen as 5% thereby the features covering all of the principle component characteristics. This is a tradeoff between the detection rate and computation time. We would like to select a suitable feature number for better discrimination and avoid confusion due to unsuitable features. For formal statistical analysis, Bartlett [15] proposed the chi-square testing method for testing the null hypothesis:

$$H_0 : \lambda_{m+1} = \cdots = \lambda_n \quad (15)$$

The statistical formula is defined as

$$\chi^2 = q \left[-\ln \left| \sum_x \right| + \sum_{j=1}^{k} \ln \lambda_j + m \ln(\ell) \right]$$
$$\approx \chi^2_{\frac{1}{2}(n-k-1)(n-k+2)} \quad (16)$$

where we choose $k = n - m$,

$$q = n - k - \frac{1}{6}\left(2m + 1 + \frac{2}{m}\right), \quad \text{and}$$

$$\ell = \frac{1}{q}\left(\text{tr}\left(\sum_x\right) - \sum_{j=1}^{k} \lambda_j\right).$$

In Bartlett chi-square testing, if the eigenvalues in the higher order are below a threshold and change smoothly (that is, the null hypothesis is accepted), it means that these higher order features are not significant for discrimination and can be ignored. These two methods were adopted in our approach for choosing the most suitable m for performance comparison. The value of "m" was selected as 26 according to Eq. (16) under a confidence level of 97.5%. Although the 26 selected features cover only 91.83% of the important attributes, the best detection rate for filled pause detection was about 76.8% and slightly better than using the features without discriminant analysis. This result was not surprising because using analyzed features is more powerful for precisely expressing and discriminating the difference between filled pauses and fluency. However, the improvement was not significant. In order to describe the specific difference more precisely, a Gaussian mixture model with multi-mixtures, i.e., many sets of analyzed features, was employed for better modeling.

4. Filled Pause Modeling Using Gaussian Mixture Model

4.1. Gaussian Mixture Model

The Gaussian mixture model (GMM) [14, 16] is a commonly used statistical model in speech and speaker recognition. In this model, the covariance matrix is usually assumed to be diagonal in applications. This assumption discards the cross-correlation between parameters and takes advantage of less computation. In speech or speaker recognition systems, the features are modeled as a class whose output probability is represented by a Gaussian mixture density. In the GMM, each Gaussian mixture with its weight of importance is used to calculate the output probability. This is because each Gaussian mixture will have a different contribution to the output probability. The framework is depicted in Fig. 8. This architecture is suitable for modeling the categorical data more precisely because it contains a large number of mixtures with their weights for describing a distribution. In this estimation process, the GMM calculates the probability of a feature $\vec{x}$ using a weighted combination of multi-variate Gaussian densities defined as

$$GMM_\lambda(\vec{x}) = \sum_{i=1}^{D} w_i N_i(\vec{x}) \quad (17)$$

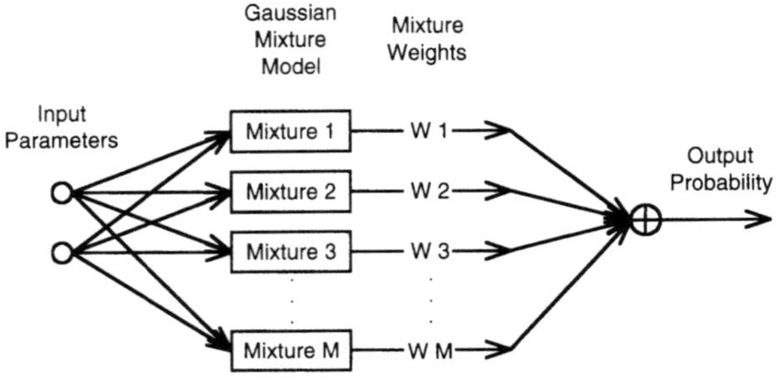

Figure 8. The framework of the GMM.

where $N_i(\vec{x})$ is the multi-variate Gaussian distribution and defined as

$$N_i(\vec{x}) = \frac{1}{\sqrt{2\pi^D |\sum|}} \exp\left(-\frac{1}{2}(\vec{x} - \vec{\mu}_i)^T \sum^{-1}(\vec{x} - \vec{\mu}_i)\right) \quad (18)$$

and w_i is the mixture weight corresponding to the i-th mixture and satisfies

$$\sum_{i=1}^{D} w_i = 1. \quad (19)$$

λ is the model and described by

$$\lambda = \left\{ w_i, \vec{\mu}_i, \sum \right\} \quad (20)$$

where $\vec{\mu}_i$ is the mean of the i-th Gaussian mixture and $\sum$ is the diagonal covariance matrix. In our approach, the 26 selected features used to analyze each filled pause in Section 2 were modeled using the GMM with 16 Gaussian mixtures using the modified k-means algorithm [20]. The weights were initially set to the ratio of the number of feature sets belonging to each Gaussian mixture with respect to the total number. They were then updated based on the gradient descent-training algorithm. There were a total of 5 filled pause GMMs for "ah," "ung," "um," "em," and "hem". Similarly, the same features for all fluent speech samples were modeled using a fluency GMM. In this modeling process the input features were fed into all Gaussian mixtures with their weights. The probability with a weighted summation of these probabilities was output using Eq. (17). The output probabilities of filled pause GMMs and fluency GMM are used to discriminate the filled pauses from fluent speech. A threshold T was used to determine the acceptance or rejection of a candidate filled pause.

4.2. Discriminative Training of Mixture Weights

In order to achieve a better discrimination between filled pauses and fluent speech, a verification function was defined to form a linear discriminator whose weights are discriminatively trained. For a given input speech x, the verification function is written as

$$V(x_t; H) = \log\left(\frac{GMM_H(x_t)}{GMM_{\bar{H}}(x_t)}\right)$$

$$= \log\left(\frac{\sum_m W_{H,m} N_{H,m}(x_t)}{\sum_m W_{\bar{H},m} N_{\bar{H},m}(x_t)}\right) \quad (21)$$

where x_t represents the t-th feature vector of the input speech, and the weight vectors $W_{H,m}$ and $W_{\bar{H},m}$ are the m-th mixture weights for the filled pause and fluency models H and $\bar{H}$, respectively. The terms $W_{H,m} N_{H,m}(x_t)$ and $W_{\bar{H},m} N_{\bar{H},m}(x_t)$ are the output probabilities of the filled pause and fluency models, respectively. A loss function was defined and minimized with respect to the weights. The loss function represents a smooth functional form of the verification score. It takes the form of a sigmoid function, written as

$$R(W_M, x_t) = \frac{1}{1 + \exp[-\eta b V(x_t; H)]} \quad (22)$$

where

$$b = \begin{cases} -1 & \text{if } x_t \in H \\ +1 & \text{if } x_t \in \bar{H} \end{cases} \quad (23)$$

$$W_M = [W_{H,M}, W_{\bar{H},M}] \quad (24)$$

and $\eta(=0.01)$ is a constant that controls the steepness of the sigmoid function.

According to the usual discriminative methodology, an optimization criterion can be defined as the minimization of the loss function and a gradient descent algorithm can interactively update the mixture weights. However, because the probability density function of x_t is not known, a gradient-based iterative procedure is used to minimize R as follows.

$$(W_M)_{n+1} = (W_M)_n - \varepsilon \nabla R((W_M)_n, X) \quad (25)$$

where ε is the updated step size and $\nabla R((W_M)_n, X)$ is the gradient of the loss function with respect to W_M evaluated by the training samples.

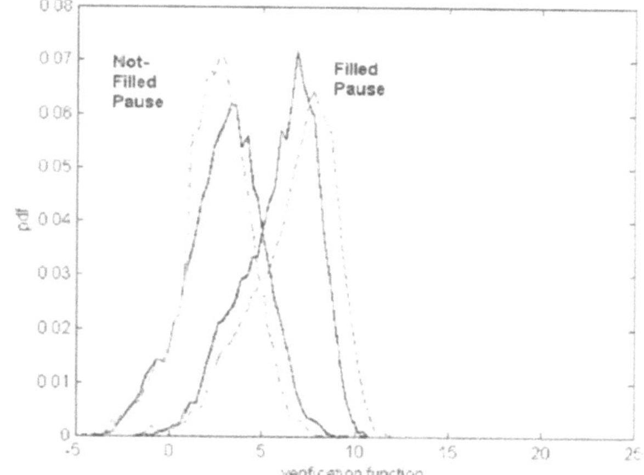

Figure 9. Probability density of the verification scores of LDA for "filled pause" and not-"filled pause." Solid line: before discriminative training. Dotted line: after discriminative training.

In our approach, using Eq. (21), we plotted the verification scores for the "filled pause" models and not-"filled pause" model. The histograms of the verification scores are shown in Fig. 9. The solid histogram on the right represents the distributions of the training samples with filled pauses. Similarly, the solid histogram on the left represents the distributions of the training samples with fluent speech. After discriminative training, the histogram of verification scores is shown in Fig. 9 with dotted line. We can see that the detection error rate for filled pauses is smaller when an optimal decision boundary is chosen after discriminative training.

5. Experiments

5.1. Database Collection

In the following experiments, a database was collected from 40 speakers. They were brought into the lab to answer questions proposed by a questioner. These conversational speech utterances were recorded and tagged into 2,160 sentences. About 12.5% of the database contains filled pauses and their corresponding sentence numbers are listed in Table 3. The most frequently used filled pause is "um" and the least frequently used filled pause is "ung." The position of the filled pause is often at the beginning of a sentence and sometime appears in the middle of a sentence. This database is divided into training and testing databases. The training database

Table 3. The number of sentences for all filled pauses in the database.

	uh	ung	um	em	hem
No. of sentences	32	24	137	48	29

Table 4. Filled pause detection rates for $m = 26, 29$, and 48 without GMM.

m	48 (Original features without KLT or LDA)	29 (KLT)	26 (KLT)	26 (LDA)
Detection rate (%)	74.0	75.6	76.8	78.1

contains 752 and 544 sentences from male and female speakers with disfluencies. The training database with filled pauses was used to train the filled pause GMMs for "ah," "ung," "um," "em," and "hem." The testing database was collected from conversational speech and consists of 756 fluent sentences and 108 sentences with filled pauses.

5.2. Experiments on Feature Selection

This experiment was conducted to select efficient discriminant features from the 48 analyzed features, which include 12 MFCCs, 12 delta MFCCs, 12 LPCs, F1, F2, F3, $MD(F1, Z1)$, $MD(F2, Z2)$, $MD(F3, Z3)$, FMR (F2, F1), FMR(F3, F1), FR(F3, F2), SE(F1), SE(F2), and SE(F3). In KLT analysis, all eigenvalues are calculated and the main component scree plot is shown in Fig. 10. If the value of m is chosen according to the traditional 5% residual error, it should be chosen as 29. If we use the Bartlett Chi-Square testing, as described in Section 2.4, the value of m should be chosen as 26 and the residual error becomes 8.17% under a confidence level of 97.5%. The detection rates for $m = 26$,

29, and 48 (original features) are listed in Table 4. The original 48 features show a good detection rate of 74%. This is because these features were chosen according to the special properties of filled pauses. In the performances of KLT for $m = 26$ and 29 in Table 4, $m = 26$ achieved the better result. In order to be clearer on comparing selection methods, the detection rates of KLT with different values of m were shown in Fig. 11. The detection rate decreased for $m \geq 26$ and the best performance happened when $m = 25$. Although the selection for $m = 26$ cannot achieve the best performance, this result shows that the KLT analysis with Bartlett Chi-Square testing can select the better discriminant features than the conventional method. As a result, considering the computation time and significant features, the acceptable value for m was 26 with a 76.8% detection rate.

In the LDA process, the projection matrix is calculated using the $S_w^{-1} S_b$ eigenvectors. We used the 26 largest eigenvalues in KLT when the samples were well separated in the 26-dimensional space. This experimental result is also shown in Table 4. The filled pause detection rate was 78.1%. An improvement of 1.3% was obtained compared to the KLT method. It is obvious that the LDA features outperformed the KLT features in detection performance under the same feature number.

5.3. Experiment on GMMs with KLT and LDA Features

In order to evaluate the detection rate for GMMs with KLT and LDA features for different values of m, a frame-based filled pause detection rate was used, defined as

$$DetectionRate = 1 - (ErrorRate_{FA} + ErrorRate_{FR})$$

(26)

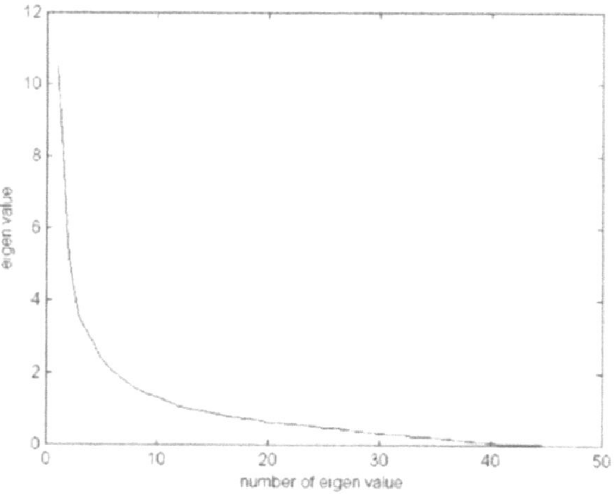

Figure 10. Main component scree plot for all eigenvalues in KLT analysis.

where the $ErrorRate_{FA}$ is the false alarm rate and $ErrorRate_{FR}$ is the false rejection rate. They are defined

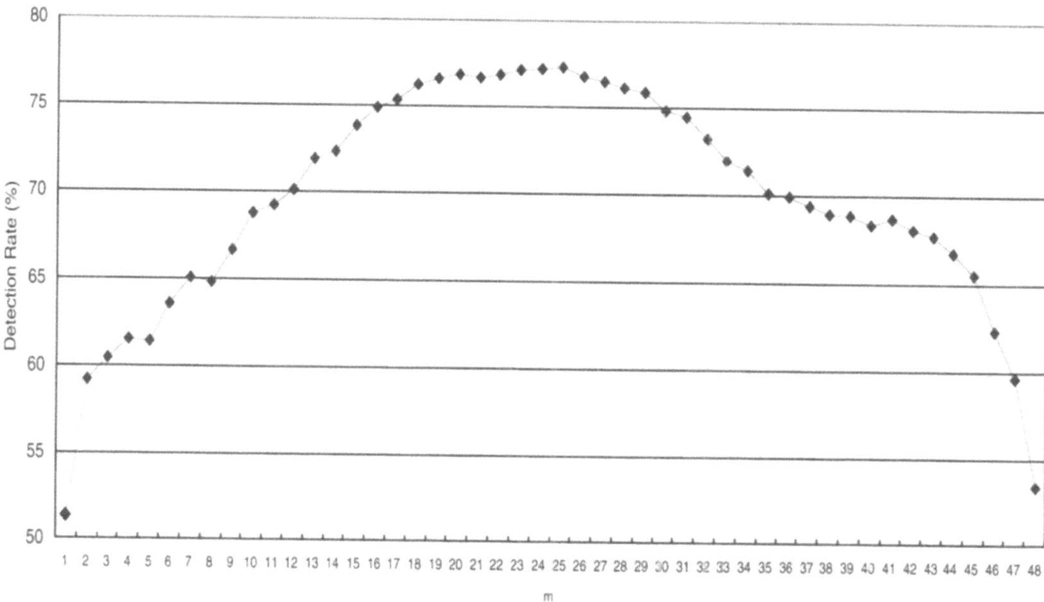

Figure 11. Detection rates of KLT with different values of m.

as follows.

$$ErrorRate_{FA} = \frac{N_{FA}}{N_{TFS}} \qquad (27)$$

$$ErrorRate_{FR} = \frac{N_{FR}}{N_{TFP}} \qquad (28)$$

where N_{TFS} is the total frame number for fluent speech patterns. N_{FA} is the frame number for fluent speech segments mis-recognized as filled pauses. Similarly, N_{TFP} is the total frame number for filled pauses, N_{FR} is the frame number of filled pauses mis-recognized as fluent speech. In the KLT analysis, the error rates for GMM with $m = 26$, 29, and 48 (original features) are shown in Fig. 12. The best detection rates for the three values of m ($m = 26$, 29, and 48) were 78.9%, 77.5%, and 75.7%, respectively. Under the same condition, the best detection rate of 79.8% for the LDA features with $m = 26$ is also shown in Fig. 12(d). This shows an improvement of 0.9% over the KLT features in detection rate and an improvement of 4.1% over the original features. It is obvious that the KLT and LDA features with $m = 26$ are more reliable for feature vector with lower dimensionality and demonstrate better performance in the experiment. The performance

improvement is because: (1) KLT and LDA do serve as a decorreclation process and diagonal covariance matrices for the GMMs are used instead of the full covariance matrices. (2) With the same amount of training data, the estimates of model parameters are more reliable for feature vector with lower dimensionality since the number of these model parameters are less.

5.4. Experiment on Discriminative GMM with KLT and LDA Features

This experiment was conducted to show the effectiveness of discriminative training and compare the detection rates of GMM and discriminative GMM with KLT and LDA features for $m = 26$ when a threshold T is chosen. Before discriminative training, an experiment on detection rate using verification score (Eq. (21)) was conducted. The result is shown in Fig. 13. The best performances for the KLT and LDA features were 81.6% and 83.2% respectively. The verification score does improve the detection performance, but this is still not the optimal result. After discriminative training, Fig. 14 shows the detection rates for discriminative GMMs. Using

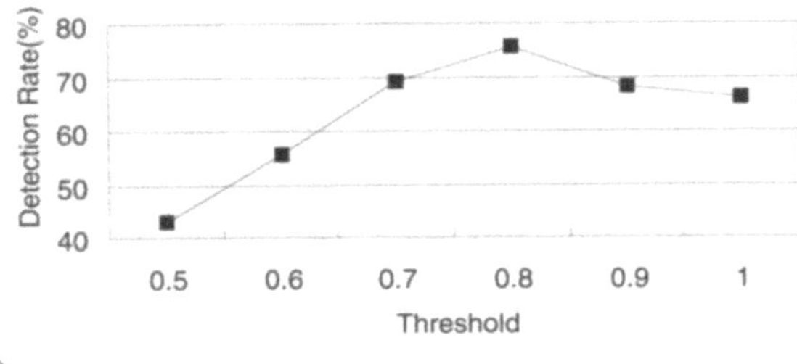

(a)

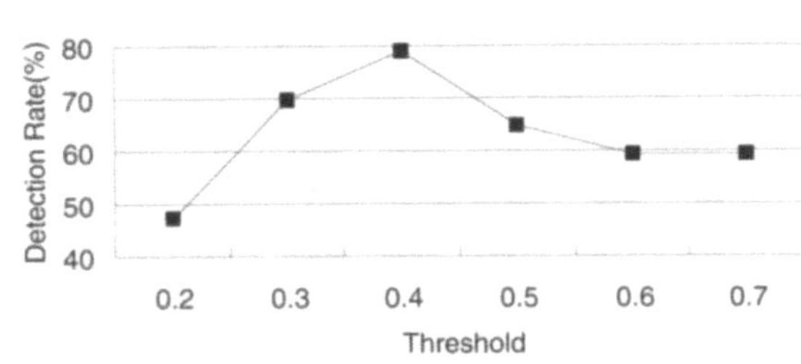

(b)

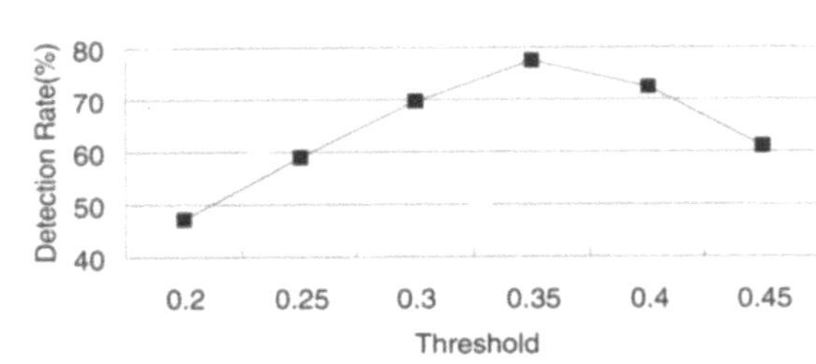

(c)

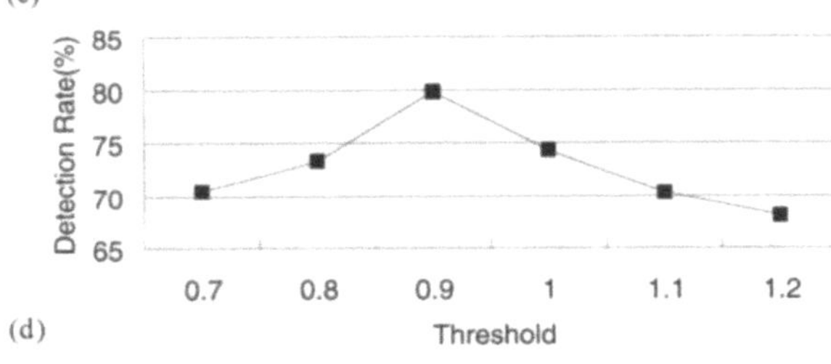

(d)

Figure 12. Detection rates using GMMs with (a) original features $m = 48$, (b) $m = 26$ (KLT), (c) $m = 29$ (KLT) and (d) $m = 26$ (LDA).

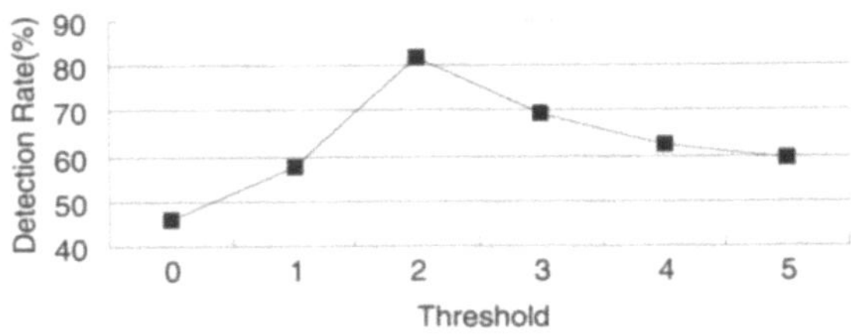

(a)

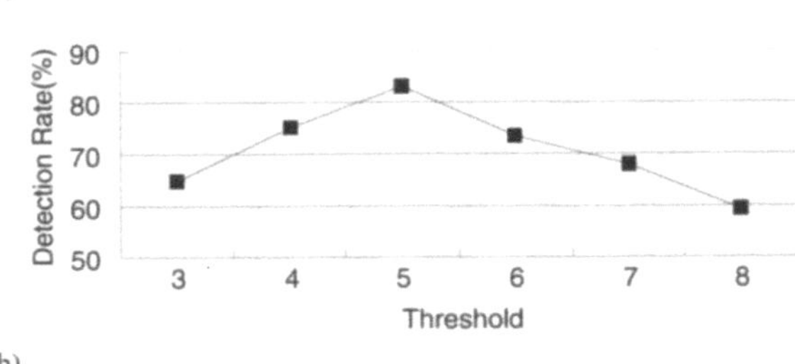

(b)

Figure 13. Detection rates using verification scores generated from GMMs with (a) KLT features ($m = 26$) and (b) LDA features ($m = 26$).

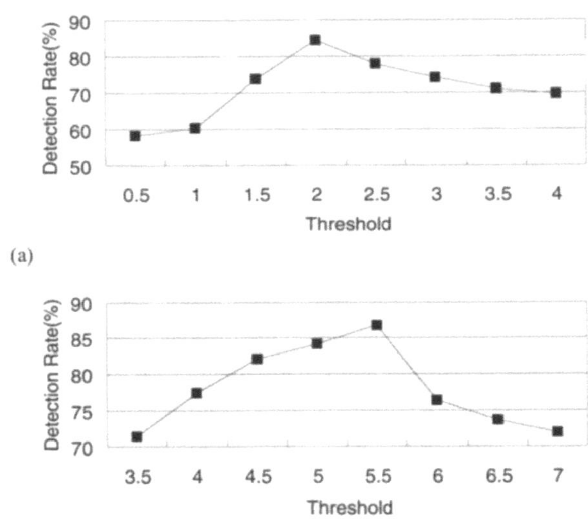

(a)

(b)

Figure 14. Detection rates using discriminative GMMs with (a) KLT features and (b) LDA features.

KLT features, when the threshold is $T = 2$, the maximum detection rate achieved 84.4%. The performance was further improved by 2.8%. For the LDA features, the maximum detection rate was 86.8% and the performance was further improved by 3.6%. The KLT features achieved an improvement of 5.5% and the LDA features achieved 7% improvement when discriminative training was applied to optimize the weights compared to the GMM without discriminative training.

6. Conclusion

In this paper, the properties of the filled pauses "ah," "ung," "um," "em," and "hem" were analyzed. Forty-eight features that describe the properties of filled pauses were first extracted and analyzed using the KLT and LDA techniques. These analyses generated the most discriminant features from 48 features automatically using the associated optimal linear projection. Finally the twenty-six discriminant features, called KLT and LDA features, were selected according to the Bartlett hypothesis testing. The KLT and LDA features were modeled using the GMM and a discriminative training methodology was employed to train the weights using a gradient-based iterative procedure. The experimental results show that the discriminative GMM achieved 84.4% and 86.8% detection rates for the KLT and LDA features, respectively. The results also show that a significant detection rate improvement was achieved using the discriminative GMM with KLT and LDA features.

References

1. W. Ward, "Understanding Spontaneous Speech: The Phoenix System," *Proc. of ICASSP-91*, 1991, pp. 365–367.
2. A. Kai and S. Nakagawa, "Investigation on Unknown Word Processing and Strategies for Spontaneous Speech Understanding," *Proc. of Eurospeech'95*, 1995, pp. 2095–2098.
3. A. Stolcke and E. Shriberg, "Statistical Language Model for Speech Disfluencies," *Proc. of ICASSP-96*, vol. 1, 1996, pp. 405–408.
4. M. Siu and M. Ostendorf, "Modeling Disfluencies in Conversation Speech," *Proc. of ICSLP-96*, vol. 1, 1996, pp. 386–389.
5. M. Siu and M. Ostendorf, "Variable N-Grams and Extensions for Conversational Speech Language Modeling," *IEEE Trans. Speech and Audio Processing*, vol. 8, no. 1, 2000, pp. 63–75.
6. L.M. Tomokiyo, "Linguistic Properties of Non-Native Speech," *Proc. of ICASSP-2000*, vol. 3, 2000, pp. 1335–1338.
7. M. Swerts, A. Wichmann, and R.J. Beun, "Filled Pauses as Markers of Discourse Structure," *Proc. ICSLP-96*, vol. 2, 1996, pp. 1033–1036.
8. D. O'Shaughnessy, "Recognition of Hesitations in Spontaneous Speech," *Proc. of ICASSP-92*, vol. 1, 1992, pp. 521–524.
9. M. Gabrea and D. O'Shaughnessy, "Detection of Filled Pauses in Spontaneous Conversation Speech," *Proc. of ICSLP-2000*, 2000.
10. G. Feng and E. Castelli, "Some Acoustic Feature of Nasal and Nasalized Vowels: A Target for Vowel Nasalization," *J. Acoust. Soc. Am.*, vol. 99, no. 6, 1996, pp. 3694–3706.
11. M.Y. Chen, "Acoustic Correlates of English and French Nasalized Vowels," *J. Acoust. Soc. Am.*, vol. 102, no. 4, 1997, pp. 2360–2370.
12. O. Fujimura, "Analysis of Nasal Consonants," *J. Acoust. Soc. Am.*, vol. 34, 1962, pp. 1865–1875.
13. D. Recasens, "Place Cues for Nasal Consonants with Special Reference to Catalan," *J. Acoust. Soc. Am.*, vol. 73, no. 4, 1983, pp. 1346–1353.
14. C.-H. Wu and G.-L. Yan, "Discriminative Disfluency Modeling for Spontaneous Speech Recognition," *EuroSpeech*, vol. 3, 2001, pp. 1955–1958.
15. W.R. Dillon and M. Goldstein, *Multivariate Analysis*, New York, U.S.A.: Wiley, 1984, pp. 44–46.
16. F. Beaufays, M. Weintraub, and K. Yochai, "Discriminative Mixture Weight Estimation for Large Gaussian Mixture Models," *Proc. of Acoustics, Speech, and Signal Processing*, vol. 1, 1999, pp. 337–340.
17. S. Ghaemmaghami, M. Deriche, and B. Boashash, "Hierarchical Approach to Formant Detection and Tracking Through Instantaneous Frequency Estimation," *Electronics Letters*, vol. 33, no. 1, 1997, pp. 17–18.
18. L.D. Swets and J. Weng, "Using Discriminant Eigenfeatures for Image Retrieval," *IEEE Trans. Pattern Analysis and Machine Intelligence*, vol. 18, no. 8, 1996.
19. A.M. Martinez and A.C. Kak, "PCA verus LDA," *IEEE Trans. Pattern Analysis and Machine Intelligence*, vol. 23, no. 2, 2001.
20. L.R. Rabiner and B.H. Juang, *Fundamentals of Speech Recognition*, New Jersey, U.S.A.: Prentice Hall, Englewood Cliffs, 1993, pp. 271–274

Chung-Hsien Wu received the B.S. degree in electronics engineering from National Chiao Tung University, Hsinchu, Taiwan, in 1981, and the M.S. and Ph.D. degrees in electrical engineering from National Cheng Kung University, Tainan, Taiwan, R.O.C., in 1987 and 1991, respectively. Since August 1991, he has been with the Department of Computer Science and Information Engineering, National Cheng Kung University, Tainan, Taiwan. He became a professor in August 1997. From 1999 to 2002, he served as the Chairman of the Department. He also worked at Massachusetts Institute of Technology Computer Science and Artificial Intelligence Laboratory, Cambridge, MA, in summer 2003 as a visiting scientist.

His research interests include speech recognition, text-to-speech, multimedia information retrieval, spoken language processing and sign language processing for hearing-impaired. Dr. Wu is a senior member of IEEE and a member of International speech communication association (ISCA) and ROCLING.
chwu@csie.ncku.edu.tw

Gwo-Lang Yan received the B.S. degree in information computer engineering from Chung-Yuan Christian University, Chung-Li, Tau-

Yuan, Taiwan, in 1995, and the M.S. degree in computer science information engineering from National Cheng Kung University in 1997. He is currently pursuing the Ph.D. degreed in the Department of computer science and information engineering, National Cheng Kung University, Tainan, Taiwan, R.O.C., and the instructor at the department of information management, Kao-Yuan Institute of Technology, Kaohsiung, Taiwan. His research interests include digital signal processing, speech recognition, keyword spotting, and natural language processing.
yangl@csie.ncku.edu.tw

Journal of VLSI Signal Processing 36, 105–116, 2004
© 2004 Kluwer Academic Publishers.

Simultaneous Recognition of Distant-Talking Speech of Multiple Talkers Based on the 3-D N-Best Search Method

PANIKOS HERACLEOUS
ATR Spoken Language Translation Research Labs, 2-2-2 Hikaridai Seika-Cho Soraku-gun, Kyoto 619-0288, Japan; Graduate School of Information Science, Nara Institute of Science and Technology, 8916-5 Ikoma Takayma, Nara 630-0101, Japan

SATOSHI NAKAMURA
ATR Spoken Language Translation Research Labs, 2-2-2 Hikaridai Seika-Cho Soraku-gun, Kyoto 619-0288, Japan

KIYOHIRO SHIKANO
Graduate School of Information Science, Nara Institute of Science and Technology, 8916-5 Ikoma Takayma, Nara 630-0101, Japan

Received October 30, 2001; Revised July 15, 2002; Accepted July 15, 2002

Abstract. This paper describes a novel method for hands-free speech recognition and in particular for simultaneous recognition of distant-talking speech of multiple sound sources (talkers or noise sources). Our method is based on the 3-D Viterbi search extended to a 3-D N-best search method to allow simultaneous speech recognition of multiple talkers. The baseline system integrates two existing technologies—3-D Viterbi search and conventional N-best search—into a complete system. However, initial evaluation of the 3-D N-best search-based system showed that new ideas were needed in order to build a system to simultaneously recognize multiple sound sources. Two factors were found to have an important role in system performance. Those two factors are the different likelihood ranges of the talkers and the direction-based separation of the hypotheses. More specifically, since we have to compare hypotheses originating from different talkers, an accurate comparison of these hypotheses cannot be made due to the different likelihood dynamic range of the talkers. Moreover, the hypotheses originated from talkers are located in different directions and therefore separating them based on their direction provides an efficient method for accurate recognition. To solve these problems, we implemented a likelihood normalization technique and a path distance-based clustering technique into the baseline 3-D N-best search-based system. The performance of our system was evaluated by experiments for recognizing the distant-talking speech of two talkers. The experiments were carried out on simulated (with only time delay) data and on reverberated (simulated and real) data. In this paper, we evaluated the proposed method in reverberant environments, and we introduced results obtained by experiments at several reverberation times and results obtained in a real environment. The experiments showed that implementing the two techniques described above produced significant improvements. Best results for simulated data were obtained by implementing the two techniques and using a microphone array composed of 32 channels. In that case in particular, the Simultaneous Word Accuracy (where both talkers are correctly recognized simultaneously) in the 'top 1' hypothesis was 72.49%, and in the 'top 3' hypotheses was 86.25%, which were very promising results.

Keywords: speech recognition, distant-talking speech, multiple sound sources, microphone array

1. Introduction

The recognition of distant-talking speech [1–4] plays an important role in any practical speech recognition system. Factors that must be considered include noisy and reverberant environments, the presence of multiple sound sources, and moving talkers. This paper deals with hands-free speech recognition, and particularly with recognizing multiple talkers. Our research is aimed at building a system for simultaneous recognition of multiple sound sources—located at fixed positions or moving—in real environments.

Most hands-free speech recognition systems are microphone array-based, since a microphone array [5–7] can obtain spatial and acoustical information from the sound source. More specifically, a microphone array can form multiple beams and can therefore be electronically steered simultaneously in multiple directions at each time frame. Use of a single microphone however provides only limited directional sensitivity and cannot be applied to localization of multiple sound sources without physical steering.

A complex problem that must be solved in systems for speech recognition system of distant-talking speech involves talker localization and speech recognition. In some approaches [8–13], the talker is first localized by using short- or long-term power and a beamformer is then steered to the hypothesized direction and recognition performed by extracting the feature vectors in this direction. However, these approaches face a serious problem in that localization of the talker is difficult under low SNR conditions. The 3-D Viterbi search method proposed by Yamada et al. [14, 15], integrates talker localization and speech recognition and performs Viterbi search in a 3-D trellis space composed of input frames, HMM states, and directions (Fig. 1). A beamformer is steered to each direction at each time frame, and this enables a locus of the sound source and a fea-

ture vector sequence to be simultaneously obtained. A 3-D Viterbi search-based system using adaptive beamforming can provide high recognition rates, but since it considers only the one best path in the 3-D trellis space, it can be applied only in the case of a single sound source.

In this paper, we propose a novel method for simultaneously recognizing multiple sound sources. This method is based on the 3-D Viterbi search extended to a 3-D N-best search method. The method performs a full search in all directions, and considers N-best word hypotheses and direction sequences. As a result, the algorithm provides an N-best list, which includes the direction sequences and phoneme sequences of multiple sound sources [16, 17]. The advantage of our proposed method is that it does not require deterministic talker localization, but uses a probabilistic approach based on the acoustic information of the input signal.

This paper describes that method along with two techniques implemented in a baseline 3-D N-best-based system. Namely, we introduce a likelihood normalization technique, which solves the problem of the different likelihood dynamic range of the talkers, and path distance-based clustering, which separates the hypotheses according to their direction.

2. 3-D Viterbi Search Extended to 3-D N-Best Search Method

The 3-D Viterbi search attempts to solve the problem of localization in the case of low SNR values, by integrating talker localization and speech recognition. The algorithm performs a Viterbi search in a 3-D trellis space and finds the optimal $(\hat{d}, \hat{q})$ path with the highest likelihood as shown in Eq. (1). In this equation, q is the state, d the direction, M the HMM model, and $\underline{X}(d)$ is the feature vector in direction d.

$$(\hat{q}, \hat{d}) = \underset{q,d}{\mathrm{argmax}}\ \Pr(\underline{X}(d)\,|\,d, q, M) \tag{1}$$

A direction sequence and a feature vector sequence can be obtained on the hypothesized path. The direction sequence corresponds to the locus of the sound source, and the feature vector sequence corresponds to the uttered speech or to other sound sources. A speech recognition system based on 3-D Viterbi search and using adaptive beamforming can provide high recognition rates and operate efficiently, even in the case of a moving talker. However, the system focuses on the

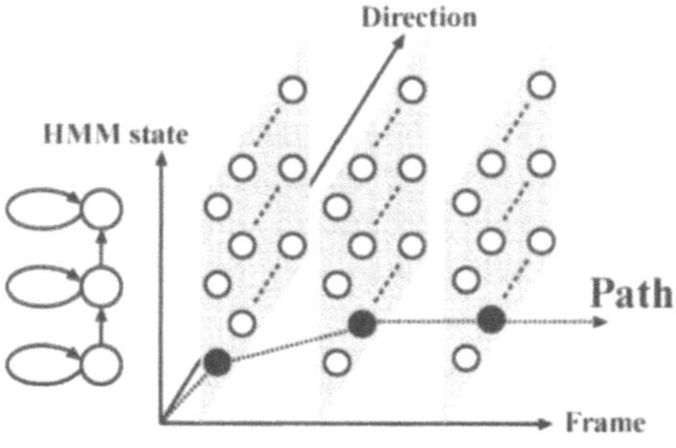

Figure 1.   3-D trellis space.

presence of only one sound source. To avoid this disadvantage, we extended the 3-D Viterbi search method to a 3-D N-best search method capable of handling multiple sound sources.

The proposed 3-D N-best search method is based on the idea that multiple sound sources can be recognized by incorporating the N-best paradigm. While the 3-D Viterbi search considers only the most likely path in a 3-D trellis space, the 3-D N-best search considers multiple hypotheses for each direction, and in this way, obtains N paths with the highest likelihoods. In a way similar to the conventional 3-D Viterbi search approach, the direction-feature vector sequences are extracted by steering the beamformer to each direction at each time frame.

The baseline 3-D N-best search is a one-pass search algorithm that performs a full search in all directions. The hypotheses arriving at a node are considered at each time frame and the N-best are found by sorting the unique ones. Equation (2) shows how the N hypotheses $\underline{\alpha}^N(q, d, t)$ with the highest likelihood are found.

$$\underline{\alpha}^N(q, d, t) = \operatorname*{sort}_{d',q'}\{\underline{\alpha}^N(q', d', t-1) + \log a_1(q', q)$$
$$+ \log a_2(d', d)\} + \log b(q, \mathbf{X}(d, t)) \tag{2}$$

The initial conditions for the above recursion are

$$\underline{\alpha}^n(1, 1, 1) = 1 \tag{3}$$

and

$$\underline{\alpha}^n(q, d, 1) = a_1(1, q)a_2(1, d)b(q, \underline{X}(d, 1)) \tag{4}$$

for $1 < q < Q$, $1 < d < D$, and $n = 0, 1 \ldots N$. Here, Q is the number of states, D the number of directions, and N indicates the N-best hypotheses.

Considering a node at time t, the overall $\underline{\alpha}^N(q', d', t-1)$ predecessor hypotheses are sorted after being added with the a_1 state and a_2 direction transition as well as the b output probabilities, so that the $\underline{\alpha}^N(q, d, t)$ N-best hypotheses can then be found. The $\underline{X}(d, t)$ is the feature vector in the direction d at time t.

The $a_1(q', q)$ state transition probability can be trained from the training data. However, automatic training of the $a_2(d', d)$ direction transition probability, which indicates how likely the talker moves, is very difficult. We therefore use a heuristic approach for estimating a_2 and it is set as follows:

$$a_2(d', d) = \begin{cases} \dfrac{1}{2\Delta d}, & |d - d'| \le \Delta d \\ 0, & |d - d'| > \Delta d \end{cases}, \tag{5}$$

where Δd is the range of the talker movements.

The initial value $a_2(1, d)$ is a uniform distribution of directions, because we assume the same equal probability for each direction.

At the last stage of the recognition system based on the 3-D N-best search, all hypotheses are sorted according to their likelihood and the 'top N' with the highest likelihood are selected. The 'top N' hypotheses are likely to contain the correct sound sources and the direction sequences are also likely to be obtained.

In the baseline system, we implemented two additional techniques as follows:

- *Likelihood normalization technique*. The N-best hypotheses are found by sorting the hypotheses that originated from different sound sources. However, different sound sources have different likelihood dynamic ranges and therefore we cannot compare them accurately. The proposed likelihood normalization technique allows comparing the hypotheses.
- *Path distance-based clustering technique*. In the case of the baseline system, there is only one N-best list that includes hypotheses that originated from different sound sources. However, if the likelihoods are high in one direction, the N-best list becomes occupied by hypotheses of the sound source located in that direction. We tried to solve this problem by implementing a path distance-based clustering technique, which separates the hypotheses according to their directions and provides an N-best list for each sound source. By finding the 'top N' for each cluster, the sound sources and their direction sequences can be obtained.

2.1. Normalization of the Likelihood of the Hypotheses

The N-best hypotheses of a (q, d) (*state, direction*) are found by sorting all the arriving hypotheses and selecting the 'top N'. However, hypotheses arriving from different directions correspond to different sound sources with different likelihood dynamic ranges. Therefore, comparing the hypotheses according to their likelihood will not be accurate. To avoid this problem, we introduce a technique for likelihood normalization.

The technique used for likelihood normalization is similar to the method proposed by Matsui et al. [18]. That method is used for speaker recognition, but can also be efficiently applied to our task. Our one-state Gaussian mixture (GM) (1 state, 64 mixtures) model is close to that proposed by Matsui et al. [18], but its objective is different. More specifically, this model runs in parallel with the other models and its accumulated likelihood is used to normalize the likelihood of the hypotheses involved. Two different techniques were implemented and compared. The two likelihood normalization techniques, L1 and L2, are the following:

- *L1 Likelihood Normalization Technique.* In the first approach, we normalize the likelihood only at the last frame. The actual likelihoods $\alpha(q, d, T)$ of every state q and direction d are normalized at the last frame T by dividing them by the likelihood $\alpha_G(d, T)$ of the one-state model. Using the logarithmic likelihoods, Eq. (6) gives the normalized likelihood $\Lambda(d, q)$.

$$\Lambda(q, d) = \alpha(q, d, T) - \alpha_G(d, T) \qquad (6)$$

- *L2 Likelihood Normalization Technique.* In this approach, the actual accumulated likelihoods $\alpha(q, d, t)$ of every state q and direction d are normalized at each time frame t by dividing them by the accumulated likelihood $\alpha_G(d, t)$ of the one-state model. Using the logarithmic likelihoods, Eq. (7) gives the normalized likelihood $\Lambda(d, q, t_f)$ at time t_f.

$$\Lambda(d, q, t_f) = \alpha(q, d, t_f) - \alpha_G(d, t_f) \qquad (7)$$

2.2. Clustering Hypotheses Using Information on Path Distances

By implementing the likelihood normalization technique, the hypotheses can be accurately compared and the performance of the system improved. However, in some cases our algorithm faces an additional problem. Namely, if the likelihoods of the hypotheses of one direction happen to be much higher than that of the other directions, the N-best list becomes occupied by hypotheses of just one direction. In this case, the algorithm fails and cannot consider all the sound sources. To solve this problem, the original 3-D N-best search was extended by implementing proposed path distance-based clustering. Here, by using information on the

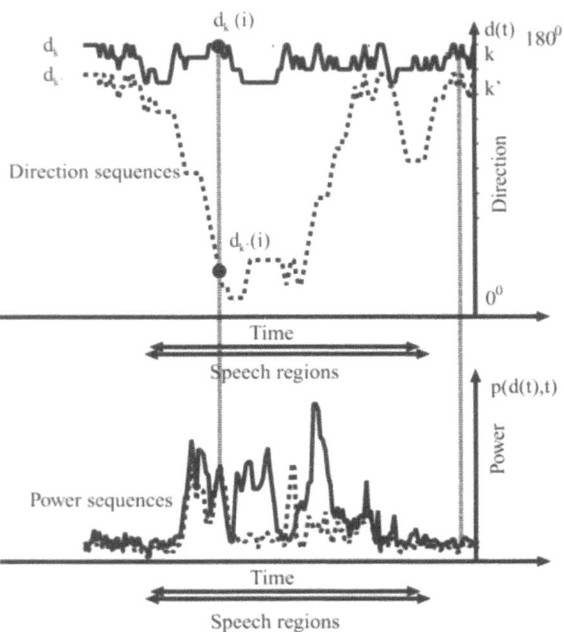

Figure 2.   Path distance of a hypothesis-pair.

provided direction sequences, the hypotheses are clustered into several clusters. Figure 2 shows the direction and power sequences of two hypotheses. In this experiment, two talkers were located at fixed positions at 10 and 170 degrees. The solid lines indicate the hypotheses that originated for the 170-degree direction, and the dotted lines indicate the hypotheses that originated from the 10-degree direction. As can be seen, in the speech region the two hypotheses are well separated based on their directions. However, in the silence region the directions of the two hypotheses appear to be similar. The reason for this is that our algorithm cannot guarantee the correct direction in the silence region. To minimize the importance of the silence region, we use the power information. Using Eq. (8), the path distance $D(k, k')$ can be calculated. Here, $D(k, k')$ is the Euclidean distance between the two direction sequences weighted by the power sequences and can be calculated as follows:

$$D(k, k') = \sum_{t=0}^{T-1} (d_k(t) - d_{k'}(t))^2 (s(d_k(t), t) \\ + s(d_{k'}(t), t)) \qquad (8)$$

In Eq. (8), T is the total number of frames, k and k' the directions at the final frames of the two hypotheses,

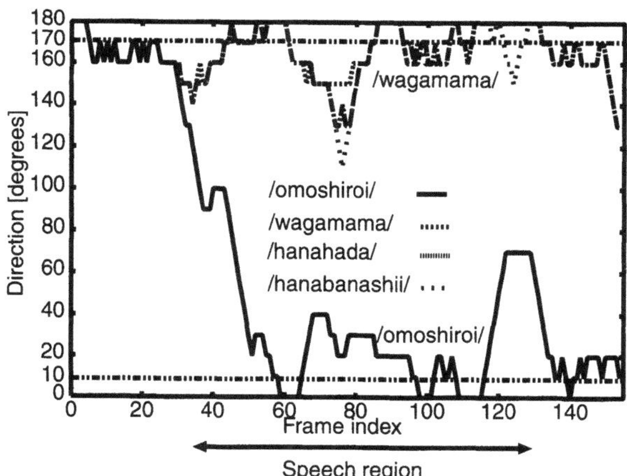

Figure 3. Results showing the transitions of hypotheses.

Table 1. Top 4 results. Sorting according to likelihood. Only one sound source was included in the *N*-best list.

Input	Speaker A /omoshiroi/	Speaker B /wagamama/
Top	Word	Likelihood
1	**/wagamama/**	−78.5579
2	/hanahada/	−78.9105
3	/hanabanashii/	−78.9776
4	/wazawaza/	−79.2003
..	..	..
7	**/omoshiroi/**	−79.5485

Table 2. Top 4 results. The hypotheses are classified using the path distance.

Input	Speaker A /omoshiroi/	Speaker B /wagamama/
Top	1st Cluster	2nd Cluster
1	**/omoshiroi/**	**/wagamama/**
2	–	/hanahada/
3	–	/hanabanashii
4	–	/wazawaza/

d_k the direction sequence ending at k, and $s(d_k)$ the power sequence corresponding to d_k. The path distance provides the measure that the clustering is based on. By using the path distance the hypotheses are classified into different clusters, which correspond to the sound sources. The number of clusters corresponds to the number of sound sources, and the sound sources can be found by selecting the 'top N' from each cluster. The sound sources directions can be obtained by examining the direction sequences of the hypotheses included in each cluster.

Figure 3 shows the transitions of four hypotheses in the case of the pronounced words /**omoshiroi**/ and /**wagamama**/. The two words were pronounced by different talkers located at respective fixed positions of 10 and 170 degrees. The aim here was to classify the hypotheses into two clusters based on the direction sequences. The words /**omoshiroi**/ and /**wagamama**/ are likely to be included in different clusters and in high orders.

Table 1 shows the results obtained for the case described in Fig. 3. The hypotheses are sorted according to likelihood and no clustering is implemented. As Table 1 shows, only the word /**wagamama**/ is included in the 'top 4' and the word /**omoshiroi**/ is of the 7th order. However, both pronounced words are likely to be of a high order. Table 2 shows the results when clustering is implemented. As can be seen, the two sound sources are included in different clusters and both are of the 1st order.

A very difficult problem that must be solved is finding the number of clusters needed for our task. In this paper, the number of clusters is pre-defined and is equal to the known number of sound sources.

3. Experiments and Results

The performance of our system was evaluated through experiments carried out under several conditions for simultaneous recognition of two talkers. We carried out experiments for simultaneously recognizing two talkers located at fixed positions, and we also considered the case of one moving and one fixed talker using simulated (with only time delay) and reverberated data. In order to find the relation between the reverberation time and accuracy, we carried out experiments at several reverberation times, using the image method to simulate those data. The performance of our system in real environments was evaluated though experiments using real data.

3.1. *Experimental Conditions*

The speech recognizer was based on a tied-mixture HMM with 256 distributions. Fifty-four context-independent phoneme models were trained with the 64-speaker ASJ speaker-independent database. The topology of the HMMs was a 3-state (emitting) left-right with no skips. The one-state GMM was also

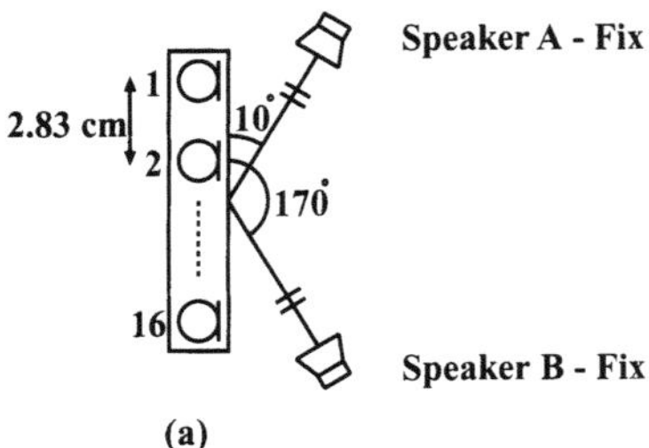

Figure 4. Source positions.

trained using the same database. The test data included 216 phoneme-balanced words of the ATR database SetA, which formed 215 word-pairs (same words were avoided). Several speaker- and words-pairs were used. More specifically, we used 10 combinations of the speakers MHT, MAU, FTK, and FSU. A total of 2150 different word-pairs were used for testing. The feature vectors were of length 33 (16 MFCC, 16 ΔMFCC, and Δpower). A linear delay-and-sum array composed of 16 channels was used for the case of a moving talker and the distance between each channel was 2.83 cm. For fixed position talkers, additional experiments were also carried out using a linear delay-and-sum array composed of 32 channels. During the search, 19 directions were considered at each time frame. The two talkers were located at fixed positions at 10 and 170 degrees as shown in Fig. 4.

3.2. *Definitions of Evaluation Measures*

In order to describe the results, two definitions are necessary. Namely, we define the Word Accuracy (WA) and Simultaneous Word Accuracy (SWA) as:

$$WA = \frac{\text{Correct Words}}{\text{Total Test Wordpairs}} \times 100\ [\%] \qquad (9)$$

$$SWA = \frac{[\text{Both Word AND Both Cluster}]\ \text{Correct}}{\text{Total Test Wordpairs}}$$
$$\times 100\ [\%] \qquad (10)$$

3.3. *Experimental Results for Fixed Position Talkers*

The first experiments were carried out using the baseline system. In the baseline system, the clustering and likelihood normalization techniques were not implemented. These experiments yielded poor results and clearly show that new ideas are needed to improve the performance of our system. The Simultaneous Word

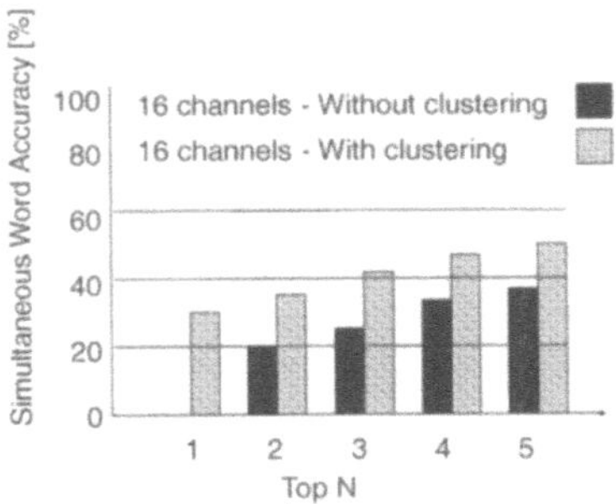

Figure 5. SWA without likelihood normalization.

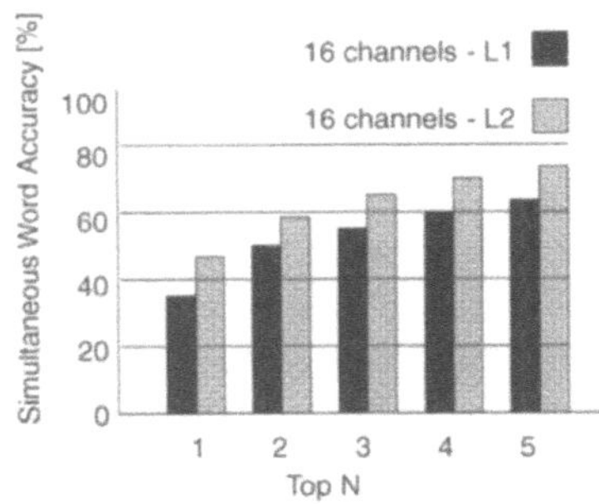

Figure 6. Results using likelihood normalization.

Accuracy (SWA) of these experiments is shown in the Fig. 5.

The performance of our system is improved by implementing the described clustering technique. As Fig. 5 shows, significant improvement can be obtained using our clustering technique. However, the expected results were not obtained.

Figure 6 shows a comparison of the two likelihood normalization techniques described in Section 2.1. As can be seen, the likelihood normalization in each time frame is more effective and therefore we selected this technique for our task.

Implementing the likelihood technique described above resulted in further improvement in accuracy. Figures 7 and 8 show the WA of the two speakers and Fig. 9 shows the SWA. As can be seen for the 'top 5', the WAs were higher than 80% and the SWAs close to 80%. The figures also show improvements achieved by implementing the likelihood normalization technique.

In the latest experiments, we used a microphone array composed of 32 channels and the two techniques were also implemented into the system. Figures 10 and 11 show the WA of the two speakers and Fig. 12 shows the SWA for a 32-channel microphone array compared

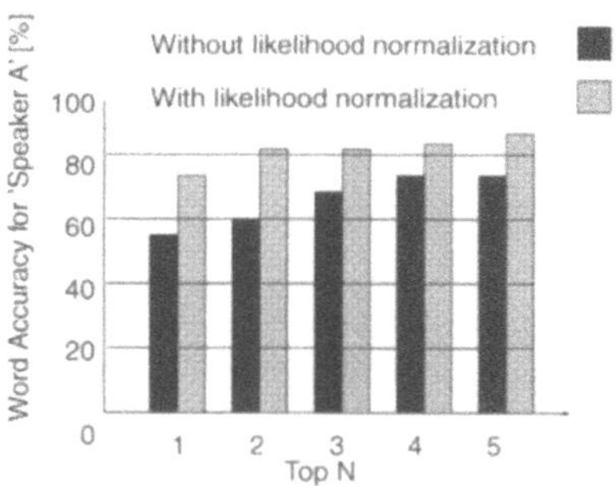

Figure 7. Word accuracy for 'Speaker A'—16 channels.

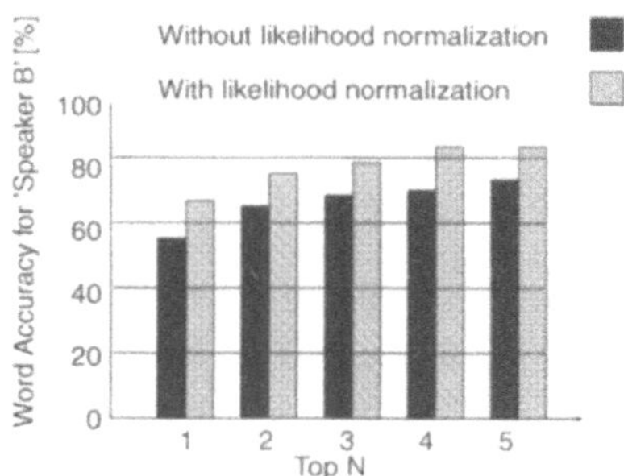

Figure 8. Word accuracy for 'Speaker B'—16 channels.

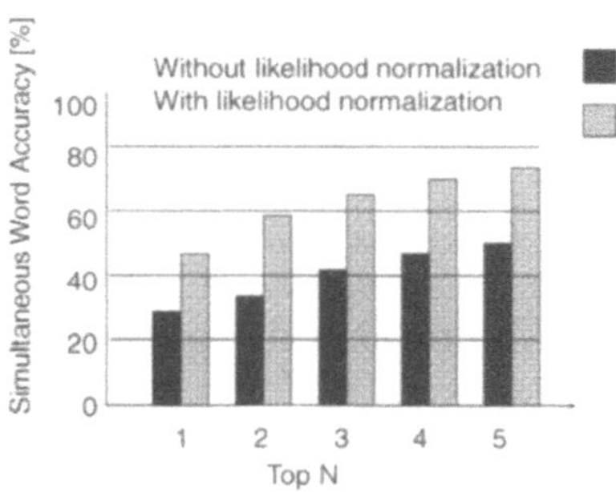

Figure 9. Simultaneous word accuracy—16 channels.

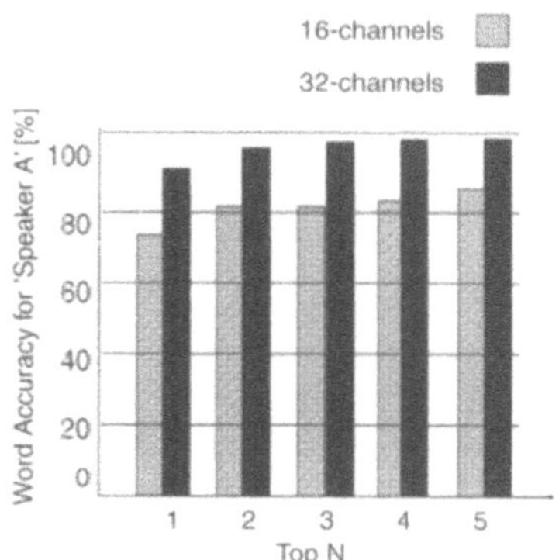

Figure 10. Word accuracy for 'Speaker A'.

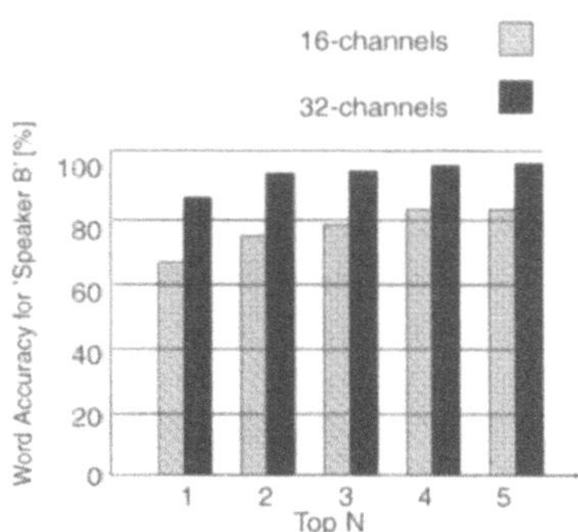

Figure 11. Word accuracy for 'Speaker B'.

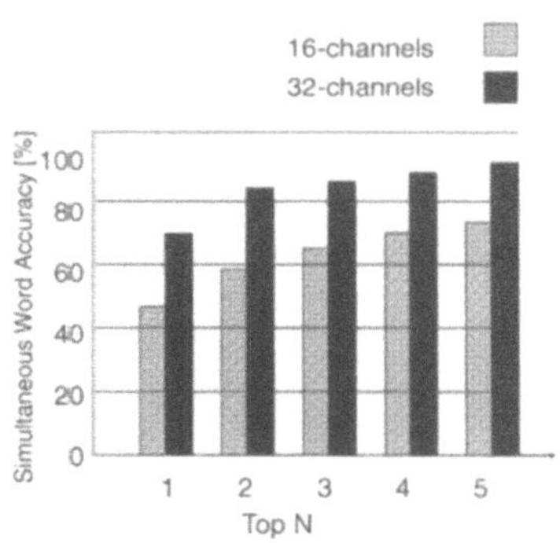

Figure 12. Simultaneous word accuracy.

with a 16-channel microphone array. As can be seen, significant improvements were achieved by increasing the number of the microphones. More specifically, for the 'top 1', the WAs were higher than 80% and the SWA higher than 70%. The reason for the improvements is that 16-channel arrays cannot form sufficiently sharp beams.

3.4. Localization Error

The localization accuracy can be described by introducing the Δ_t localization error—in degrees—defined as follows:

$$\Delta_t = |d_t - d_t'|(s(d_t, t) + s(d_t', t)) \qquad (11)$$

Here, d_t is the correct direction, and d_t' is the hypothesized direction in frame t. To minimize the effect of

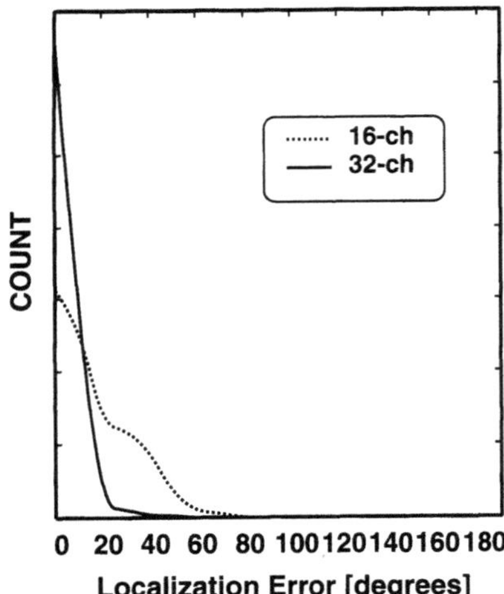

Figure 13. Localization error.

the silence region, we also introduce the $s(d, t)$ which gives the power value of the input speech in direction d at time t. Figure 13 shows the histogram of the localization error in the case of talkers located at fixed positions and using 16 and 32 channels. In constructing the $h(\Delta_t^n)$ histogram, we consider only the 'top 1' of each cluster and we calculate the Δ_t for each frame t. As the results show, when using a microphone array composed of 32 channels our system localizes the talkers with very high accuracy. In contrast, 16 channels appear to provide poor localization accuracy. This is the main reason that the WA in the case of 32 channels are much higher compared to those of 16 channels. Figure 14 shows the directivity pat-

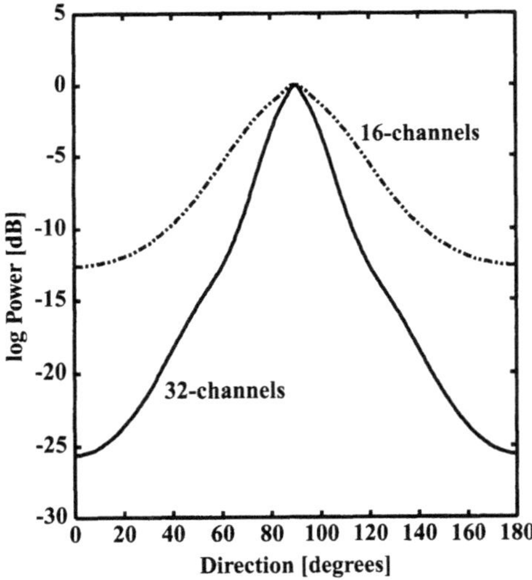

Figure 14. Directivity patterns.

terns of the two microphone array, when the sound source is located at 90 degrees. As can be seen, the 32-channel microphone array forms a much sharper beam.

3.5. Comparing a 3-D N-Best Search-Based System with a CSP (Cross-Power Spectrum Phase)-Based System

We carried out an experiment to compare the performance of a 3-D N-best search-based system with a conventional localization method-based system. More specifically, we carried out an experiment using talkers FSU and FTK located at fixed positions at 30 and 150 degrees. A 32-channel linear delay-and-sum microphone was used. The CSP method [8, 9] was selected for comparison with our 3-D N-best search method. In the CSP, the DOA (Direction Of Arrival) of the speech signal can be obtained by estimating the delay in arrival between two channels. After the sound source was localized, a 2-D Viterbi search was performed. Table 3 shows results that were obtained. As can be seen the 3-D N-best search-based system shows better performance.

3.6. Experimental Results for a Moving Talker

To make our system more versatile and eliminate any restrictions on its application, we also considered the case of a moving talker. In this experiment, one of the two talkers was located at a fixed position at 10 degrees, and the other one moved from 0 to 180 degrees uttering a word as shown in Fig. 15. A microphone array composed of 16 channels was used. The distance between the channels was 2.83 cm. The distance between the microphone array and the moving talker was 2 m. The talker moved at a speed of about 80cm per second. In this experiment simulated data (only with time delay) was used. By using the 3-D N-best search-based system, the WA for the moving talker in the 'top 5' was

Table 3. Comparing 3-D N-best search and CSP method.

Localize method	Search method	Source	Direction		SWA (%)
			A	B	
CSP	2-D Viterbi	2	Fix-30	Fix-150	63.72
	3-D N-best Search	2	Fix-30	Fix-150	73.68

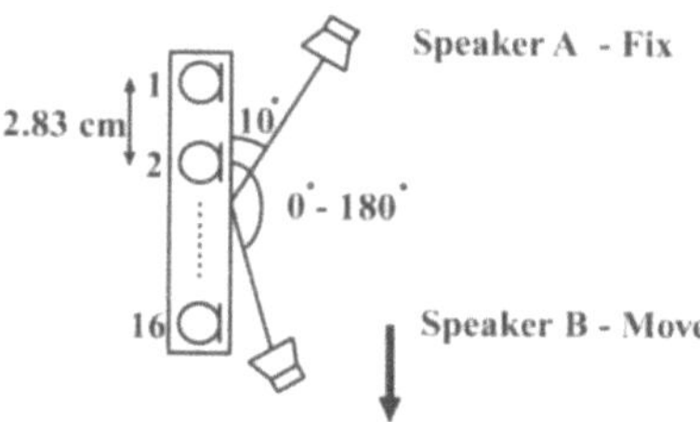

Figure 15.   Source positions in the moving talker's case.

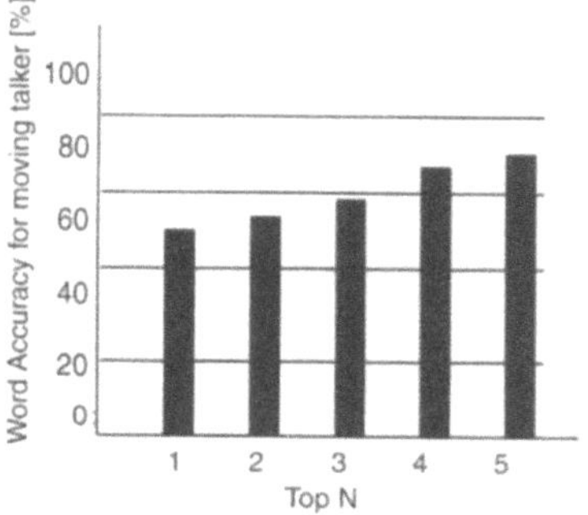

Figure 16.   Word accuracy for moving talker.

72.01%. Figure 16 shows the results obtained. Compared with the 2-D Viterbi search, the 3-D N-best search has the additional advantage of being able to recognize a moving talker in an unknown direction. Performance of a speech recognition systems using conventional localization method, such as CSP, depends greatly on the accurate localization of the talker. With these systems, however, accurate localization appears to be very difficult to achieve with a moving talker. The results showed that our proposed 3-D N-best-based method performs efficiently, even when one of the two talkers is moving.

3.7.   *Experimental Results Using Simulated Reverberated Data*

In these experiments we used the *Image Method* [19] to simulate reverberated data and evaluate the performance of our 3-D N-best search-based system in real environments. The impulse responses provided by the image method are convoluted with the clean speech in order to obtain reverberated speech.

Figure 17 describes the arrangement of the experiments. Two sound sources were located at fixed positions at 10 and 170 degrees. The MHT and FSU speakers were located at 10 and 170 degrees, respectively. A

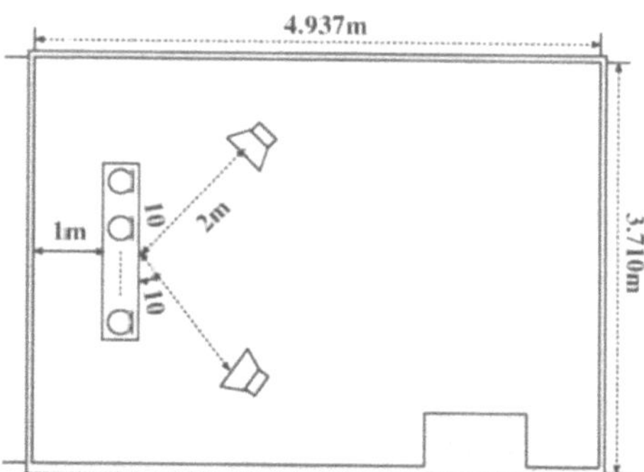

Figure 17.   Experiment arrangement for reverberant environment.

linear microphone array composed of 32 channels was used. The distance between the channels was 2.83 cm. The distance between the sound sources and the microphone array was 2 m. The microphone array is located in the center of the room 1 m away from the wall.

The reverberation time was measured by using square integration. A reverberation attenuation curve can be obtained based on the impulse response. The reverberation time is considered to be the time where the attenuation becomes 60 dB.

We carried out experiments at different reverberation times. More specifically, we carried out speaker-independent isolated word recognition experiments for reverberation times of 162-, 200- and 240-ms. Figures 18–20 show the achieved results. These results show differences in word accuracy for 'Speaker A' and 'Speaker B'. The reason for this was the quality of speech of the two talkers. More specifically, the MHT speaker appears to be recognized with higher accuracy than the FSU speaker.

The results that were obtained reflect the effect of reverberation on speech recognition. The performance of our system degrees under a reverberant environment. The relative decreases in Simultaneous Word Accuracy are 27%, 37.5% and 52% for the 162-, 200- and 240-ms reverberation times. However, these are not bad results. At 162-ms reverberation time in particular, the Simultaneous Word Accuracy in the 'top 1' hypothesis was 52.56% and in the 'top 3' hypotheses was 73.02%, which were reasonable results. Our system can therefore be considered relatively robust with regard to reverberation.

Reverberation is a difficult problem for hands-free speech recognition systems. A great number of researchers are working in this field to develop methods for de-reverberation. In this paper, we did

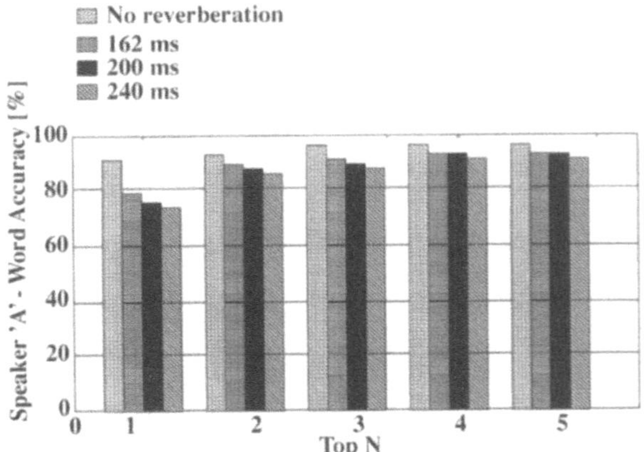

Figure 18.   Speaker 'A' WA.

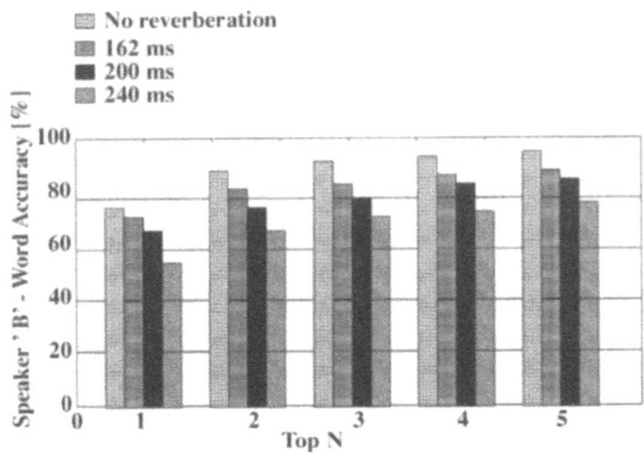

Figure 19.   Speaker 'B' WA.

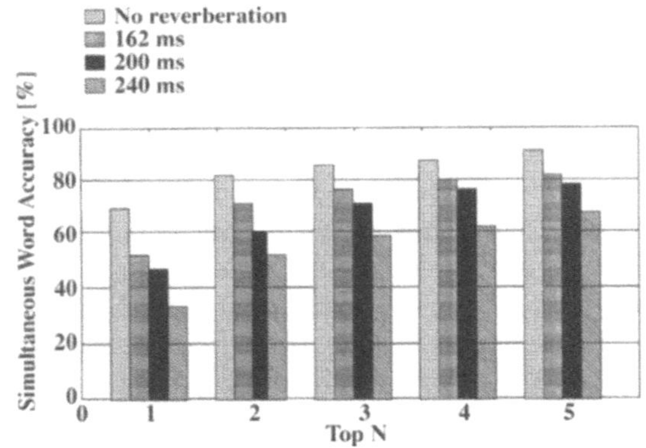

Figure 20.   Simultaneous word accuracy.

not implement any de-reverberation technique, other than delay-and-sum beamforming. However, delay-and-sum beamforming is not efficient for performing de-reverberation.

3.8.   *Experiments Using Real Reverberated Data*

This section describes experiments carried out for simultaneous recognition of distant-talking speech in real environments. Figure 17 shows the experiments conditions. The MHT and FSU talkers were located at respective fixed positions at 10 and 170 degrees. The speech data were played back through loudspeakers. A linear microphone array composed of 32 channels was used and the distance between each channel was 2.80 cm. The distance between the loudspeakers and the microphone array was 2 m. The reverberation time ($T_{[60]}$) in the experimental room was 280-ms.

Figures 21–23 show the results obtained compared with the results using the image method. Although the performance of our system decreased in real environments, the results we achieved are comparable with those when simulated reverberated data were used. However, in the case of using real data, we should also consider the presence of ambient noise and a longer reverberation time. We can therefore conclude that the

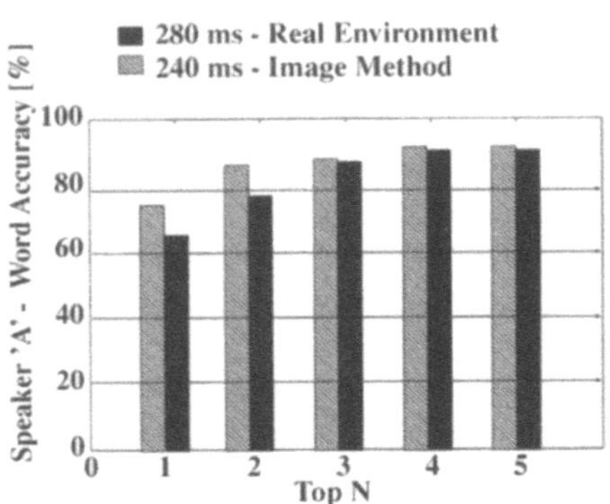

Figure 21.   Speaker 'A' WA.

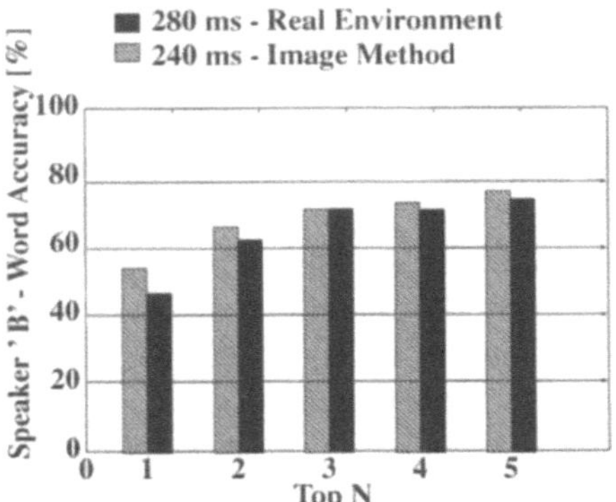

Figure 22.   Speaker 'B' WA.

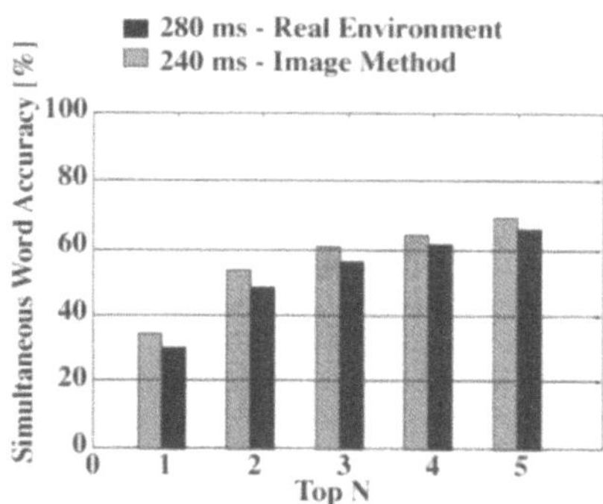

Figure 23.   Simultaneous word accuracy.

results obtained using real data will be lower than those using simulated data.

4.   Conclusion

In this paper we proposed 3-D N-best search, which is a novel method for simultaneously recognizing multiple sound sources. We described two problems faced by the baseline 3-D N-best search and we introduced possible solutions. Namely, a likelihood normalization technique and a clustering technique were implemented into the baseline 3-D N-best search-based system, which succesfully improved the recognition rates. The performance of our system was evaluated through experiments both on simulated and real data for two talkers both located at fixed positions, and with one moving and one fixed talker. The results confirmed that improvements were obtained by implementing these two techniques. The performance of our system was further improved by increasing the number of microphones in the microphone array. More specifically, significant improvements were obtained by increasing the number of microphones from 16 to 32. In the final case, we achieved a Simultaneous Word Accuracy of nearly 72%. The performance of our system decreases in reverberant environments. Additional tecniques seem to be necessary for de-reverberation. We plan to investigate this problem in future work. Our algorithm requires a large computational load. So we also plan to implement new ideas to speed-up the system.

References

1. S. Nakamura, T. Yamada, T. Takiguchi, and K. Shikano, "Hands-Free Speech Recognition by a Microphone Array and HMM Composition," in *Proc. of International Workshop on Human Interface Technology*, 1995, pp. 33–38,

2. M. Inoue, S. Nakamura, T. Yamada, and K. Shikano, "Microphone Array Design Measures for Hands-Free Speech Recognition," in *Proc. of European Conference on Speech Communication and Technology*, 1997, pp. 331–334.

3. S. Nakamura, T. Yamada, P. Heracleous, and K. Shikano, "Recognition of Distant-Talking Speech Based on 3-D Trellis Search Using a Microphone Array and Adaptive Beamforming, in *Proc of Workshop on Robust Methods for Speech Recognition*, 1999, pp. 219–222.

4. T. Takiguchi, S. Nakamura, and K. Shikano, "Speech Recognition for a Distant Moving Speaker Based on HMM Composition and Separation," in *Proc. of IEEE International Conference on Acoustics. Speech, and Signal Processing*, 2000, pp. 1403–1406.

5. G.W. Elko, *Superdirectional Microphone Arrays*, Acoustic Signal Processing for Telecommunications, Kluwer Academic Publishers, 2000.

6. D.H. Johnson and D.E. Dudgeon, *Array Signal Processing*, Concepts and Techniques PTR Prentice-Hall, Inc., 1993.

7. J.L. Flanagan. D.A. Berkley, G.W. Elko, J.E. West, and M.M. Sondhi, "Autodirective Microphone Systems," *Acoustica* vol. 75, 1991.

8. M. Omologo and P. Svaizer, "Talker Localization and Speech Recognition Using a Microphone Array and a Cross-Power Spectrum Phase Analysis", *Pro-c. ICSLP*, 1994, pp. 1243–1246.

9. M. Omologo and P. Svaizer, "Acoustic Source Location in Noisy and Reverberant Environment Using CSP Analysis," *Proc. ICASSP*, 1996, pp. 921–924.

10. T. Yamada, S. Nakamura, and K. Shikano, "Robust Speech Recognition with Speaker Localization by a Microphone Array," *Proc. ICSLP*, 1996, pp. 1317–1320.

11. P. Svaizer, M. Matassoni, and M. Omologo, "Acoustic Source Location in a Three-Dimensional Space Using Crosspower Spectrum Phase," *Proc. ICASSP*, 1997, pp. 231–234.

12. T. Hughes, H. Kim, J. DiBiase, and H. Silverman, "Using a Real Time, Tracking Microphone Array as Input to an HMM Speech Recognizer," *Proc. ICASSP*, 1998, pp. 249–252.

13. T. Nishiura, T. Yamada, S. Nakamura, and K. Shikano, "Localization of Multiple Sound Sources Based on a CSP Analysis with a Microphone Array," *Proc. ICASSP*, 2000, pp. 1053–1056.

14. T. Yamada, S. Nakamura, K. Shikano, "Hands-free Speech Recognition Based on 3-D Viterbi Search Using a Microphone Array," *Proc. ICASSP*, 1998, pp. 245–248.

15. T. Yamada, S. Nakamura, and K. Shikano, "An Effect of Adaptive Beamforming on Hands-free Speech Recognition Based on 3-D Viterbi Search," *Proc. ICSLP*, 1998, pp. 381–384.

16. P. Heraclecus, T. Yamada, S. Nakamura, and K. Shikano, "Simultaneous Recognition of Multiple Sound Sources based on 3-D N-best Search using Microphone Array," *Proc. Eurospeech99*. 1999, pp. 69–72.

17. P. Heracleous, S. Nakamura, and K. Shikano, "Multiple Sound Sources Recognition by a Microphone Array-based 3-D N-best Search with Likelihood Normalization," in *Proc. International Workshop on Hands-free Speech Communication*, 2001, pp 103–107.

18. T. Matsui and S. Furui, "Likelihood Normalization for Speaker Verification using a Phoneme- and Speaker-independent

Model," *Speech Communication*, vol. 17, 1995, pp. 109–116.

19. J.B. Allen and D.A. Berkley. Image Method for Efficiently Simulating Small-Room Acoustics. *Journal of Acoustical Society of America*, vol. 65, no 4, 1979, pp. 943–950.

Panikos Heracleous was born in Cyprus on May 22, 1966. He received the M.Sc. degree in Communication Engineering from Technical University of Budapest, Hungary in 1992, and the Dr. Eng. degree from Nara Institute of Science and Technology, Japan in 2002. During 1997–1998 he was a research student at Graduate School of Information Science, Nara Institute of Science and Technology, Japan. During 1998–2001 he was a Ph.D. student at Graduate School of Information Science, Nara Institute of Science and Technology, Japan. During 2000–2001 he was student intern, and intern researcher in ATR Spoken Language Translation Research Laboratories, Japan. Since 2001 he has been at KDDI R&D Laboratories Inc, Japan where he is a research engineer. His research interests include keyword spotting algorithms, robust speech recognition, hands-free speech recognition, and microphone array processing. He is member of the Acoustical Society of Japan (ASJ), Institute of Electrical and Electronics Engineers (IEICE), and the Institute of Electrical and Electronics Engineers (IEEE).
panikos@is.aist-nara.ac.jp

Satoshi Nakamura received the B.S. degree in electronics engineering from Kyoto Institute of Technology in 1981 and the Ph.D. degree in information science from Kyoto University in 1992. Between 1981–1986 and 1990–1993, he worked with the Central Research Laboratory, Sharp Corporation, Nara, Japan, where he was engaged in speech recognition research. During 1986–1989, he was a researcher of the speech processing department at ATR Interpreting Telephony Research Laboratories. From 1994–2000, he was an associate professor of the graduate school of information science, Nara Institute of Science and Technology, Japan. In 1996, he was a visiting research professor of the CAIP center of Rutgers, the state university of New Jersey, USA. He is currently the head of Department 1 in ATR Spoken Language Translation Laboratories, Japan. He also serves as a guest professor for Toyohashi University of Technology and Ritsumeikan University from April 2002. His current research interests include speech recognition, speech translation, spoken dialogue systems, stochastic modeling of speech, and microphone arrays. He received the Awaya Award from the Acoustical Society of Japan in 1992, and the Interaction2001 best paper award from the Information Processing Society of Japan in 2001. He is a member of the Acoustical Society of Japan, Institute of Electrical and Electronics Engineers (IEICE), Information Processing Society of Japan, and IEEE. He is currently a member of the Speech Technical Committee of the IEEE Signal Processing Society and an editor for the Journal of the IEICE Information and System Society.
nakamura@slt.atr.co.jp

Kiyohiro Shikano received the B.S., M.S., and Ph.D. degrees in electrical engineering from Nagoya University in 1970, 1972, and 1980, respectively. He is currently a professor of Nara Institute of Science and Technology (NAIST), where he is directing speech and acoustics laboratory. His major research areas are speech recognition, multi-modal dialog system, speech enhancement, adaptive microphone array, and acoustic field reproduction. From 1972 to 1993, he had been working at NTT Laboratories, where he had been engaged in speech recognition research. During 1986–1990, he was the Head of Speech Processing Department at ATR Interpreting Telephony Research Laboratories. During 1984–1986, he was a visiting scientist in Carnegie Mellon University. He received the Yonezawa Prize from IEICE in 1975, the Signal Processing Society 1990 Senior Award from IEEE in 1991, the Technical Development Award from ASJ in 1994, IPSJ Yamashita SIG Research Award in 2000, and Paper Award from the Virtual Reality Society of Japan in 2001. He is a member of the Institute of Electronics, Information and Communication Engineers of Japan (IEICE), Information Processing Society of Japan, the Acoustical Society of Japan (ASJ), Japan VR Society, the Institute of Electrical and Electronics, Engineers (IEEE), and International Speech Communication Society.
shikano@is.aist-nara.ac.jp

Journal of VLSI Signal Processing 36, 117–124, 2004

Multi-Modal Speech Recognition Using Optical-Flow Analysis for Lip Images

SATOSHI TAMURA, KOJI IWANO AND SADAOKI FURUI
*Department of Computer Science, Graduate School of Information Science and Engineering,
Tokyo Institute of Technology, 2-12-1 Ookayama, Meguro-ku, Tokyo 152-8552, Japan*

Received October 30, 2001; Revised July 26, 2002; Accepted August 13, 2002

Abstract. This paper proposes a multi-modal speech recognition method using optical-flow analysis for lip images. Optical flow is defined as the distribution of apparent velocities in the movement of brightness patterns in an image. Since the optical flow is computed without extracting the speaker's lip contours and location, robust visual features can be obtained for lip movements. Our method calculates two kinds of visual feature sets in each frame. The first feature set consists of variances of vertical and horizontal components of optical-flow vectors. These are useful for estimating silence/pause periods in noisy conditions since they represent movement of the speaker's mouth. The second feature set consists of maximum and minimum values of integral of the optical flow. These are expected to be more effective than the first set since this feature set has not only silence/pause information but also open/close status of the speaker's mouth. Each of the feature sets is combined with an acoustic feature set in the framework of HMM-based recognition. Triphone HMMs are trained using the combined parameter sets extracted from clean speech data. Noise-corrupted speech recognition experiments have been carried out using audio-visual data from 11 male speakers uttering connected digits. The following improvements of digit accuracy over the audio-only recognition scheme have been achieved when the visual information was used only for silence HMM: 4% at SNR = 5 dB and 13% at SNR = 10 dB using the integral information of optical flow as the visual feature set.

Keywords: multi-modal speech recognition, optical flow, robust to noise, speaker independent

1. Introduction

Automatic Speech Recognition (ASR) systems are expected to play important roles in an advanced multimedia society with user-friendly human-machine interfaces, such as an ubiquitous computing environment [1, 2]. Although high recognition accuracy can be obtained for clean speech using the state-of-the-art technology even if the vocabulary size is large, the accuracy largely decreases in noisy environments. Increasing the robustness to noisy environments is one of the most important issues of ASR.

Multi-modal speech recognition, in which acoustic features and other information are used jointly, has been investigated and found to increase robustness and thus improve the accuracy of ASR. Most of the multi-modal methods use visual features, typically lip information, in addition to the acoustic features [3–9]. By using the visual information, acoustically similar sounds, such as nasal sounds: /n/, /m/, and /ng/, become easier to recognize [7, 10]. In most of the studies, a lip contour is extracted from an image by mouth tracking and pattern matching techniques. Since it is not easy to determine a mouth location and extract a lip shape, lip marking is often needed to ensure robust extraction of visual features.

Mase and Pentland reported their lip-reading system for recognizing connected English digits using an optical-flow analysis [10]. Optical flow is defined as the distribution of apparent velocities in the movement of brightness patterns in an image [11]. Various advantages exist in using the optical flow for audio-visual bimodal speech recognition, and they stated the following advantages of the optical-flow method for lip reading: First, the optical-flow vectors can be calculated without using any prior knowledge about the shape of the object. Thus the visual features can be detected without extracting the lip locations and contours. Second, from a viewpoint of human vision, it is more reasonable to use lip motion for lip-reading systems rather than using a lip shape. Third, it is easier to find word boundaries by looking for the instance of zero velocity than using shape analysis. Fourth, the visual features for recognition are independent of the speaker's mouth shape or, in some cases, beard. They set four windows around the speaker's mouth in the face image, and extracted horizontal and vertical mean in each window. They tested their method based on pattern matching, and obtained a high performance.

In this paper, we propose using the optical-flow analysis for robust audio-visual bimodal speech recognition. Acoustic and visual features are combined in each frame to construct the audio-visual features to be used for model training and recognition. We describe the recognition results of noise-corrupted speech and compare the results with the audio-only recognition method. This paper is organized as follows: In Section 2 the principle of the optical-flow method is explained. Our audio-visual bimodal speech recognition system is described in Section 3. Experimental setup and results are shown in Sections 4 and 5. Finally we conclude our research and describe our future works in Section 6.

2.   Optical-Flow Analysis

Optical flow is the distribution of apparent velocities in the movement of brightness patterns in an image [11]. We use the Horn-Schunck algorithm [11], the most typical method among a family of the optical-flow analysis. This algorithm has an advantage that it needs no characteristic point in contrast with pattern-matching-based algorithms, and that it requires only two images in processing. In this method, we assume that brightness at every point is constant, for a short time, during movement. With this assumption, the time derivative of the brightness is zero:

$$\frac{dE}{dt} \simeq \frac{\partial E}{\partial x} \cdot \frac{dx}{dt} + \frac{\partial E}{\partial y} \cdot \frac{dy}{dt} + \frac{\partial E}{\partial t} = 0 \quad (1)$$

-where $E(x, y, t)$ is a brightness of a point (x, y) in

an image at time t. In this Eq. (1), if we let

$$u = \frac{dx}{dt} \quad \text{and} \quad v = \frac{dy}{dt} \quad (2)$$

then the following constraint is obtained:

$$E_x \cdot u + E_y \cdot v + E_t = 0 \quad (3)$$

Here $u(x, y)$ denotes a horizontal element of optical flow at a point (x, y), and $v(x, y)$ denotes a vertical element. Since we cannot determine $u(x, y)$ and $v(x, y)$ using the Eq. (3) only, we incorporate another restraint called "smoothness constraint":

$$\iint \left\{ \left(u_x^2 + u_y^2 \right) + \left(v_x^2 + v_y^2 \right) \right\} dx\, dy \to \min \quad (4)$$

which means that the square of the magnitude of the gradient of $u(x, y)$ and $v(x, y)$ at every point must be minimized. In other words, the optical-flow velocity becomes smooth between every neighboring two pixels in an image. The optical-flow vectors $u(x, y)$ and $v(x, y)$ are computed under these two constraints (3) and (4) by an iterative technique using the average of optical-flow velocities estimated over neighboring pixels. An example of the optical-flow analysis is shown in Fig. 1. The left image (a) is extracted from a video sequence at a certain time, and the right image (b) is the next picture. An image of optical-flow velocities computed from these images is shown in (c).

3.   Audio-Visual Bimodal Speech Recognition System

3.1.   Feature Extraction

Figure 2 shows structure of our audio-visual bimodal speech recognition system. Speech signals are recorded at a 16 kHz sampling rate, and a speech frame with the length of 25 ms is extracted at every 10 ms. Each frame is converted into 39-dimensional acoustic parameters: 12-dimensional mel-frequency cepstral coefficients, normalized log energy, and their first and second order derivatives. A video stream is captured with the frame rate of 15 frames/sec and the resolution size of 720×480. Before computing the optical flow, the resolution is reduced to 180×120 keeping the aspect ratio, and the image is transformed into gray scale. Low-pass filtering (smoothing) and low-level noise addition are applied in order to increase the precision

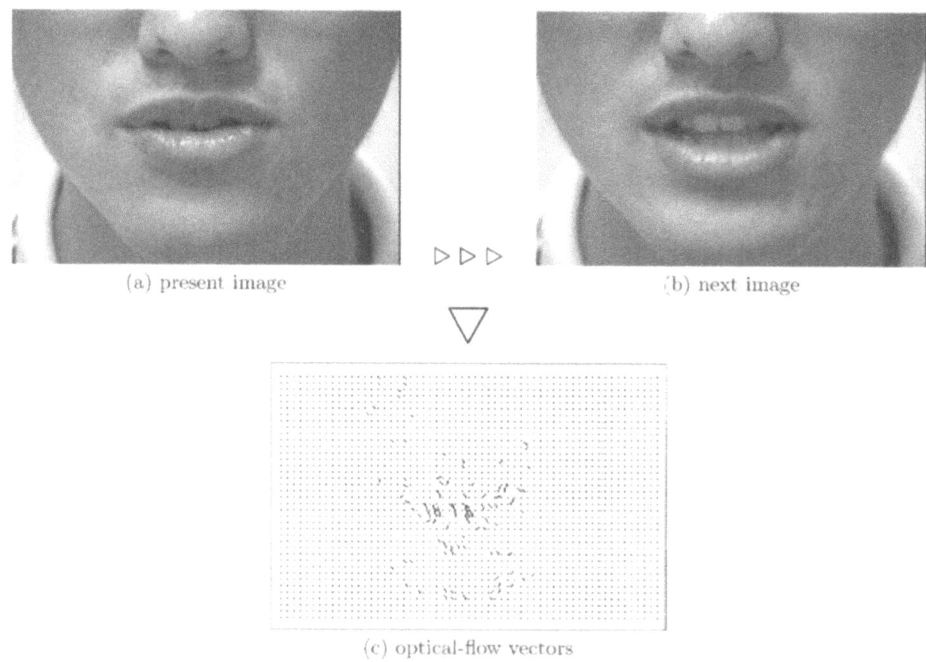

Figure 1.   A example of optical-flow analysis.

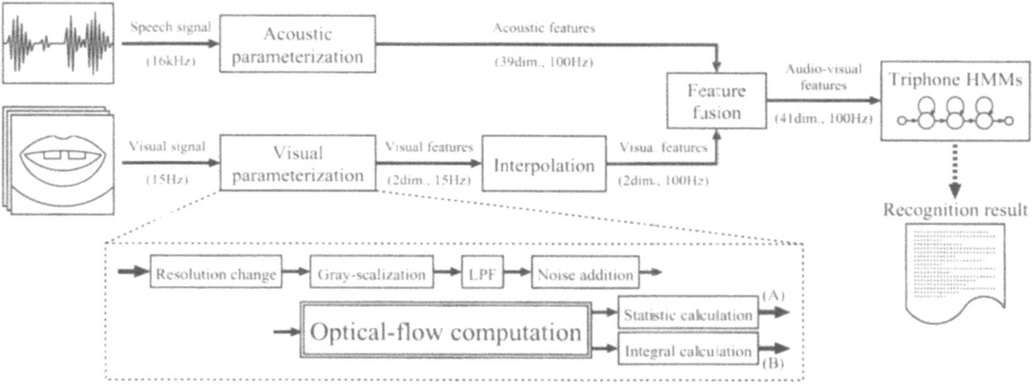

Figure 2.   Audio-visual bimodal speech recognition system.

of the optical flow, since the optical-flow vectors cannot be properly determined if brightness derivatives are almost zero or extremely large, i.e. in flat areas or near boundaries of objects. Then the optical flow is computed from a pair of consecutive images with 5 iterations.

We extract and compare two kinds of visual features from the optical-flow analysis. The first feature set (A) consists of the horizontal and vertical variances of whole optical-flow vector components. This feature reflects whether speaker's mouth is moving or not. It becomes zero when the mouth is not moving, whereas it has some positive value while speaking, since the optical-flow vectors exist only around moving points. Therefore, this feature set is expected to be effective in detecting silence or pause periods.

The second feature set (B) consists of the maximum and minimum values of the integral of the optical-flow vectors. The 3-D images of integral results are shown in Fig. 3. When a speaker is not speaking, the surface is almost flat as shown in the left image (a). When the speaker's mouth is opening, optical-flow vectors point

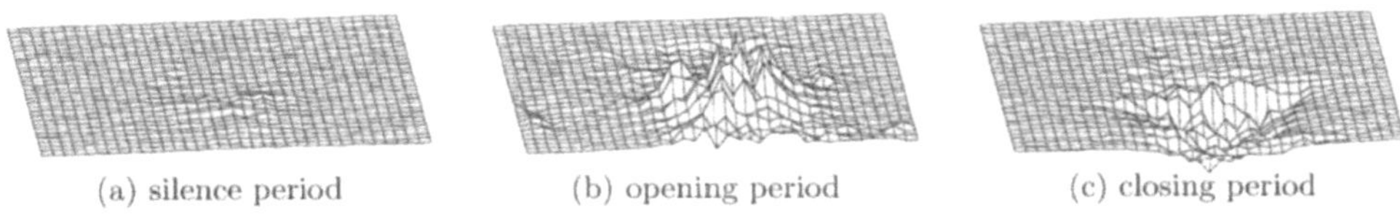

(a) silence period   (b) opening period   (c) closing period

Figure 3.   Examples of optical-flow integral result.

in diffusing directions around the mouth shape. As a result, a mountain-like surface is created as shown in the center image (b), and it produces the maximum value. When the mouth is closing, converging vectors of optical flow occur around the lip contour. Then the integral operation produces a dip in the mouth area, as shown in the right image (c), and the minimum value is observed. Therefore, the second feature contains not only silence information but also open/close information of the mouth.

Note that the two kinds of visual features (A) and (B), each composing a 2-dimensional vector, are independent of the lip shape and are not used together. Always either of the features is used in the following process.

3.2.   Feature Fusion

There are two data fusion methods used in multi-modal speech recognition systems: "feature fusion" and "decision fusion" [12]. "Feature fusion" is the method in which features are extracted from the raw data and subsequently combined. "Decision fusion" is performed at the most advanced stage of processing and involves combining the decision of all different classifies making independent decisions. As for bimodal speech recognition, in the feature fusion case, once the acoustic and visual features are combined into a single audio-visual vector, the same model training and recognition methods as normal ASR systems can be used. In contrast, in the decision fusion case, the training and recognition are performed for audio and visual modalities independently before the final decision is made. The feature fusion method usually needs more training data than the decision fusion method.

Since we have sufficient data for training, we decided to employ the feature fusion method in our experiments. In order to apply this method, acoustic and visual frame rates need to be synchronized. Hence visual features are interpolated from 15 frames/sec to 100 frames/sec with 3-degree spline function before the feature fusion method is applied. As a result of fu-

sion, 41-dimensional audio-visual vectors consisting of the 39-dimensional audio part and the 2-dimensional visual part are generated.

3.3.   Modeling

A set of triphone Hidden Markov Models (HMMs) having 3 states and 2 mixtures in each state is used as a model in our system. All the triphone HMMs are trained using the 41-dimensional audio-visual features with the EM algorithm. In the recognition stage of our bimodal ASR system, the feature vector is divided into two streams: the 39-dimensional acoustic feature vector (audio stream) and the 2-dimensional visual feature vector (visual stream). The observation probability $b_j(O_{AV})$ of generating audio-visual features O_{AV} is given by the following Eq. (5):

$$b_j(O_{AV}) = b_{A_j}(O_A)^{\lambda_A} \times b_{V_j}(O_V)^{\lambda_V} \qquad (5)$$

where $b_{A_j}(O_A)$ and $b_{V_j}(O_V)$ are probabilities of generating the acoustic vector O_A and the visual vector O_V in a state j respectively, λ_A and λ_V are weighting factors for the audio stream and the visual stream, respectively. Properly setting λ_A and λ_V is important especially in noisy conditions. Under a high noise condition, λ_A should be relatively small and λ_V should be relatively high, since confidence of the audio information is decreased, whereas visual information is not influenced at all. By properly controlling the weighting factors according to the noise condition, improvement of the recognition accuracy compared with the audio-only ASR is expected.

4.   Experimental Setup

4.1.   Database

An audio-visual speech database was collected in a clean/quiet condition from 11 male speakers, each uttering 250 sequences of connected digits. Each sequence consisted of 2-6 digits, such as "39", "714886", and "87549", with an average of 4 digits. The total duration of our database was approximately 2.5 hours.

A microphone and a DV camera were located in front of a speaker roughly 1 m away and the speaker's face was illuminated by ceiling lights. We recorded images around speaker's mouth as shown in (a) and (b) in Fig. 1. The zooming and positioning of the images were set manually before recording, and they were not changed throughout the recording for each subject. Since the features used in our method are independent of the lip location in the image, no detection technique was used. Several speakers had a beard or mustache, but no speaker had lip marking. The audio and visual signals were recorded simultaneously in the DV tape.

4.2. Training and Recognition

Experiments were conducted using the leave-one-out method: data from one speaker were used for testing while data from other 10 speakers were used for training. This process was rotated for all possible combinations.

As the first step of experiments, this paper focuses on the usefulness of lip moving information for silence detection in ASR. Therefore, we controlled the stream weight parameters, λ_A and λ_V, only for the silence HMM under the following constraint:

$$\lambda_A + \lambda_V = 1, \quad \lambda_A \geq 0, \quad \lambda_V \geq 0 \qquad (6)$$

We fixed λ_A and λ_V at 1.0 and 0.0 respectively for any other triphone HMM, that is, visual features were considered only for the silence triphone in the recognition process.

All the HMMs were built using only clean audio-visual signals. To test the robustness of our method in noisy conditions, white noise was added to the audio signal in the audio-visual data for testing at SNR levels of 5 dB, 10 dB, and 20 dB. Clean audio-visual signal was also used for comparison of the recognition results.

5. Experimental Results

Figure 4 shows the digit recognition results at 5 dB, 10 dB, and 20 dB SNR level conditions, and the clean condition.

The horizontal axis indicates the audio stream weight λ_A, and vertical axis indicates the percentage of digit

recognition accuracy computed as follows:

$$Accuracy = \frac{N - D - S - I}{N} \times 100 \ [\%] \qquad (7)$$

where N is the total number of uttered digits, D is "deletion error" that is the number of missed digits, S is the "substitution error" that is the number of substituted digits, and I is the "insertion error". The dotted line indicates the accuracy of an audio-only ASR system as the baseline. Two solid lines with points 'o' and '•' indicate performance of bimodal ASR systems which use the visual features (A) and (B), respectively. Since the audio-only ASR is not influenced by the stream weight, the dotted line is flat. Alternatively, the accuracy of the bimodal ASR system varies as a function of the stream weight controlling the balance of the two streams: an audio stream and a visual stream. The optimum weight maximizing the accuracy varies according to the SNR value. These results show that our bimodal ASR system achieves better performance than an audio-only ASR in noisy environments. At SNR = 5 dB, roughly 4–5% improvement is observed. Approximately 12% and 13% improvements have been achieved with visual information (A) and (B) respectively at SNR = 10 dB. In the SNR = 20 dB and clean conditions, little improvement is observed, since the baseline is so high that the accuracy is almost saturated without adding visual features.

Comparing the maximum digit accuracy of visual feature sets (A) and (B), both of them show almost the same performance. As we described, the visual feature set (B) contains further information comparing with (A). The reason why the performance of (B) is not more superior than (A) is considered that still we cannot use all information of (B) because we control λ_A and λ_V only for silence HMM. However, the performance of (B) is better than (A) totally shown in Fig. 4. This means that visual information (B) is more useful for real applications since the accuracy is not much influenced by the stream weight factor λ_A.

In our database, each speaker uttered 250 sequences of connected digits with approximately 237 silence periods on average between them. In the above evaluation, the silence periods were removed from the speech signals for recognition, and silences inserted within digit sequence durations as recognition results were not counted as errors. However, in actual applications, it is important that digits are not inserted in silence periods and silences should not be inserted within digit sequences. Therefore, we have conducted an additional

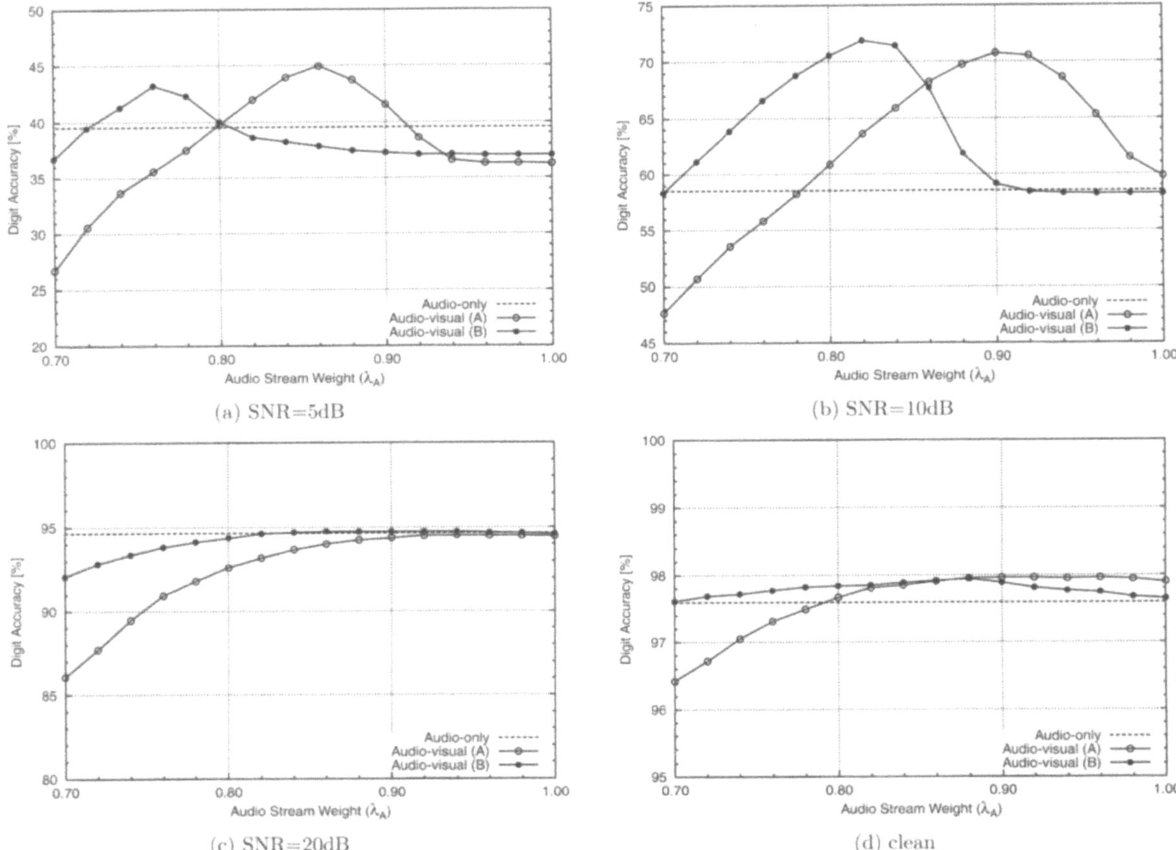

(a) SNR=5dB

(b) SNR=10dB

(c) SNR=20dB

(d) clean

Figure 4.   Word accuracy in various SNR levels of white noise.

evaluation in which silence insertion within speech periods and substitution of silences by digits as well as digit insertion within silences were counted as errors. Table 1 shows the comparison between results

Table 1.   The number of deletion (Del), substitution (Sub), and insertion (Ins) errors.

	Del	Sub	Ins
Audio only			
With silence	14.36	25.01	4.12
Without silence	9.64	18.90	12.91
Audio-visual (A)			
With silence	15.41	9.61	1.81
Without silence	14.56	12.92	1.74
Audio-visual (B)			
With silence	14.47	9.52	2.35
Without silence	14.01	11.83	2.27

of evaluations "with silence" and "without silence" for both audio-only and audio-visual methods at the best stream weighting condition when SNR = 10 dB: $\lambda_A = 0.90$ for audio-visual parameter set (A) and $\lambda_A = 0.82$ for (B). The table shows the number of errors within 100 utterances per speaker. The difference between the results with/without silence is clear for substitution and insertion errors in the audio-only condition, whereas there are few differences in the audio-visual condition. This means that silence periods are much more correctly detected by the audio-visual method than the audio-only method.

The visual features used in this paper are effective mainly to detect silences and they are too simple to recognize digits by themselves. Therefore, if SNR is very low and the audio signal is severely distorted, the improvement by combining the visual features is limited. This is probably because the improvement shown in Fig. 4 by combining the visual features at SNR = 5 dB is smaller than that at SNR = 10 dB.

6. Conclusions and Future Works

In this paper we proposed an audio-visual bimodal speech recognition scheme using optical-flow analysis of lip movements. Two kinds of visual feature sets were extracted and each of them was combined with traditional acoustic features using the feature-fusion method. The audio-visual features were used in both training and testing, in which the stream weight was varied only for the silence HMM and fixed for other triphone HMMs. The experimental results show that the proposed method achieves approximately 13% improvement at SNR = 10 dB and 4% at SNR = 5 dB when using the integral information of optical-flow vectors as visual features. The visual features are significantly useful for detecting silence periods and reducing digit insertion errors in pause/silence periods.

Our future works include: (1) investigation of more robust and informative visual parameters, such as features including direction and amount of lip movements, (2) optimization of the stream weight for each triphone HMM to improve the performance by applying the maximum likelihood method or other algorithms, and (3) collecting a new database in real environments, such as a car navigation environment in a driving automobile, for increasing the robustness in extracting the audio-visual information.

Acknowledgments

This research has been conducted in cooperation with NTT DoCoMo. The authors wish to express thanks for their support.

References

1. S. Furui, "Speech Recognition Technology in the Ubiquitous/ Wearable Computing Environment," in *Proc. ICASSP2000*, vol. 6, 2000, pp. 3735–3738.
2. S. Furui, K. Iwano, C. Hori, T. Shinozaki, Y. Saito, and S. Tamura, "Ubiquitous Speech Processing," in *Proc. ICASSP2001*, vol. 1, 2001, pp. 13–16.
3. K. Iwano, S. Tamura, and S. Furui, "Bimodal Speech Recognition Using Lip Movement Measured by Optical-Flow Analysis," in *Proc. HSC2001*, 2001, pp. 187–190.
4. S. Nakamura, H. Ito, and K. Shikano, "Stream Weight Optimization of Speech and Lip Image Sequence for Audio-Visual Speech Recognition," in *Proc. ICSLP2000*, vol. 3, 2000, pp. 20–24.
5. C. Miyajima, K. Tokuda, and T. Kitamura, "Audio-Visual Speech Recognition Using MCE-Based HMMs and Model-Dependent Stream weights," in *Proc. ICSLP2000*, vol. 2, 2000, pp. 1023–1026.
6. Y. Zhang, S. Levinson, and T. Huang, "Speaker Independent Audio-Visual Speech Recognition," in *Proc. ICME2000*, TP8-1, 2000.
7. S. Basu, C. Neti, N. Rajput, A. Senior, L. Subramaniam, and A. Verma, "Audio-Visual Large Vocabulary Continuous Speech Recognition in the Broadcast Domain," in *Proc. MMSP'99*, 1999, pp. 475–481.
8. G. Potamianos, E. Cosatto, H.P. Gref, and D.B. Roe, "Speaker Independent Audio-Visual Database for Bimodal ASR," in *Proc. AVSP'97*, 1997, pp. 65–68.
9. C. Bregler and Y. Konig, "Eigenlips" for Robust Speech Recognition," in *Proc. ICASSP'94*, vol. 2, 1994, pp. 669–672.
10. K. Mase and A. Pentland, "Automatic Lipreading by Optical-Flow Analysis," *Trans. Systems and Computers in Japan*, vol. 22, no. 6, 1991, pp. 67–76.
11. B.K.P. Horn and B.G. Schunck, "Determining Optical Flow," *Artificial Intelligence*, vol. 17, nos. 1-3, 1981, pp. 185–203.
12. D.L. Hall, *Mathematical Techniques in Multisensor Data Fusion* Artech House, Boston, 1992.

Satoshi Tamura is a Ph.D. candidate at Tokyo Institute of Technology (TIT). He received the M.E. degree in information science and engineering from TIT in 2002. His research interests are speech information processing, especially multi-modal audio-visual speech recognition. He is a member of the Acoustical Society of Japan (ASJ). tamura@furui.cs.titech.ac.jp

Koji Iwano received a B.E. degree in information and communication engineering in 1995, and a M.E. and Ph.D. degree in information engineering respectively in 1997 and 2000 from the University of Tokyo. He is currently an Assistant Professor at Tokyo Institute of Technology, Department of Computer Science. His research interests are speech information processing, such as speech recognition, speaker recognition, and speech synthesis. He is a member of the International Speech Communication Association (ISCA), the Institute

of Electronics, Information and Communication Engineers (IEICE), the Information Processing Society of Japan (IPSJ), and the Acoustical Society of Japan (ASJ).
iwano@furui.cs.titech.ac.jp

Sadaoki Furui is currently a Professor at Tokyo Institute of Technology, Department of Computer Science. He is engaged in a wide range of research on speech analysis, speech recognition, speaker recognition, speech synthesis, and multimodal human-computer interaction and has authored or coauthored over 400 published articles. He is a Fellow of the IEEE, the Acoustical Society of America and the Institute of Electronics, Information and Communication Engineers of Japan (IEICE). He served as President of the Acoustical Society of Japan (ASJ) from 2001 to 2003 and he is now President of the International Speech Communication Association (ISCA) and the Permanent Council for International Conferences on Spoken Language Processing (PC-ICSLP). He is a Board of Governor of the IEEE Signal Processing Society. He has served as Editor-in-Chief of the Transaction of the IEICE and an Editor-in-Chief of Speech Communication. He has received the Yonezawa Prize, the Paper Award and the Achievement Award from the IEICE (1975, 88, 93, 2003), and the Sato Paper Award from the ASJ (1985, 87). He has received the Senior Award from the IEEE ASSP Society (1989) and the Achievement Award from the Minister of Science and Technology, Japan (1989). He has received the Book Award from the IEICE (1990). In 1993 he served as an IEEE SPS Distinguished Lecturer.
furui@furui.cs.titech.ac.jp

Journal of VLSI Signal Processing 36, 125–139, 2004
© 2004 Kluwer Academic Publishers.

Speech Enhancement Using Perceptual Wavelet Packet Decomposition and Teager Energy Operator

SHI-HUANG CHEN

Department of Computer Science and Information Engineering, Shu-Te University No. 59, Hun Shan Rd. Yen Chau, Kaohsiung Count, 824 Taiwan, Republic of China

JHING-FA WANG

Department of Electrical Engineering, National Cheng Kung University, 1 University Road, Tainan, Taiwan 701, Republic of China

Received October 30, 2001; Revised September 13, 2002; Accepted November 18, 2002

Abstract. It has been shown in the literature that the perceptual wavelet packet decomposition (PWPD) and the Teager energy operator (TEO) are useful for various speech processing systems and speech enhancement applications, respectively. By the use of the PWPD and the TEO, this paper presents an improved wavelet-based speech enhancement method. The main advantage of the proposed method is that the over thresholding of speech segments which is usually occurred in conventional wavelet-based speech enhancement schemes can be avoided. As a consequence, the enhanced speech quality of the proposed method can be increased substantially from those of conventional approaches. In addition, the proposed method does not require a complicated estimation of the noise level or any knowledge of the SNR. Using speech signals corrupted by additive and real noises, experimental results demonstrate that the speech enhancement method presented in this paper is capable of outperforming conventional noise cancellation schemes.

Keywords: speech enhancement, perceptual wavelet packet decomposition (PWPD), teager energy operator (TEO), time-adaptive thresholding (TAT)

1. Introduction

In practice, a speech signal is usually corrupted by background noise in its acquisition or transmission. Hence the speech enhancement plays an important role in the recognition or compression of speech signals at increased system performances [1, 2]. For example, the speech enhancement scheme can be incorporated as a post-processing step in voice communication systems in order to reduce listener's fatigue as well as improving the perceptual quality of speech [3]. Traditionally, the speech enhancement is achieved by the use of linear processing techniques such as Wiener filtering, linear prediction, spectral subtraction, and etc. [1–6]. Recently, a vast literature has emerged on speech

enhancement using nonlinear techniques like wavelet. The wavelet transform is acknowledged as a powerful tool for the nonlinear filtering of signals corrupted by noises. Now wavelet transform has been widely used in image denoising and speech enhancement [7–11].

One popular technique for wavelet-based signal enhancement is the wavelet shrinkage algorithm which was proposed by Donoho and Johnstone [12, 13]. This algorithm is based on thresholding the wavelet coefficients of the noise-corrupted signal and can be easily applied to speech enhancement. It has been shown [7, 10] that such a wavelet thresholding scheme has a better performance of speech enhancement than those of traditional methods. If one lets the noisy speech signal to be $x(n) = s(n) + g(n)$, $n = 1, \ldots, N$, where $s(n)$ and $g(n)$ represent a clean speech signal and a Gaussian white noise, respectively. Then the wavelet-based speech enhancement algorithm can be summarized in the following three steps:

(a) Decompose the noisy speech signal into wavelet coefficients.
(b) Employ a threshold method to the wavelet coefficients obtained in (a).
(c) Synthesize these thresholded wavelet coefficients obtained in (b) to achieve the enhanced speech signal.

For removing additive Gaussian white noise in the signal, namely $x(n)$, Donoho and Johnstone [12, 13] proposed a time-constant threshold value λ expressed as

$$\lambda = \sigma \sqrt{2 \log(N)} \qquad (1)$$

where $\sigma = \text{MAD}/0.6745$ is the estimated noise level. The median absolution deviation (MAD) is determined in the first decomposition stage of the wavelet transform. For the wavelet packet case, the threshold becomes

$$\lambda = \sigma \sqrt{2 \log(N \log_2 N)}. \qquad (2)$$

Although the above algorithm is very simple and of satisfying performance, there still has two important issues which need to be improved. First, due to the time variability and non-stationarity of speech signals, the utilization of a time-constant threshold value will lead to over thresholding of speech signals. In other words, the use of an identical threshold value will not only suppress unwanted noises but also some speech segments liked unvoiced ones [10]. Consequently, the perceptual quality of the enhanced speech is degraded. This drawback will become serious when the speech signal is just contaminated by slight noises. Second, a better wavelet decomposition algorithm for speech signal needs to be obtained. It follows from [14] that the wavelet shrinkage algorithm is practically restricted by the extent to which the wavelet transform decomposes the unknown signal and noise into several significant coefficients. Therefore, utilizing an inappropriate wavelet decomposition will limit the performance of the wavelet-based speech enhancement method.

In order to overcome the above two problems, this paper presents an improved wavelet-based speech enhancement method using the perceptual wavelet packet decomposition (PWPD) and the Teager energy operator (TEO). In this improved method, the TEO and the PWPD are applied to develop a time-variance threshold value and an appropriate wavelet decomposition algorithm for speech enhancement, respectively. Many schemes of the PWPD are proposed for various speech analysis applications such as speech enhancement [8, 9], speech coding [9], speech recognition [16], and speaker identification [17]. In contrast to the conventional wavelet packet decomposition, the psychoacoustic models of human hearing are embedded in the decomposition rule of the PWPD. The bandwidths of the subbands generated from PWPD are designed to match the auditory critical bands as close as possible. It is shown in the literature [8, 9, 16–18] that the PWPD has better performances than the conventional wavelet transform on dealing with speech or audio signals. The TEO, which is a powerful nonlinear operator proposed by Kaiser [19, 20], is used successfully in the speech recognition system [21, 22] and is proposed for speech enhancement recently [10]. In [10], the TEO is developed to generate a time-adaptive threshold (TAT) value with masks construction. The main feature of the TAT is that its threshold value is time adapted in function of speech components. Hence the speech enhancement method proposed in this paper combines the advantages of the PWPD and the TEO in order to increase the perceptual speech quality after enhancement processing. With the techniques of PWPD and TEO, the over thresholding of speech segments that usually occurred in conventional wavelet-based speech enhancement schemes can be avoided. As a consequence, the enhanced speech quality of the proposed method can be increased substantially from those of conventional approaches. In addition, this improved method does not require a complicated estimation of the noise level or any knowledge of the SNR.

The system implementation for the proposed speech enhancement method using PWPD and TEO is shown in Figure 1. The PWPD is first applied to the noisy

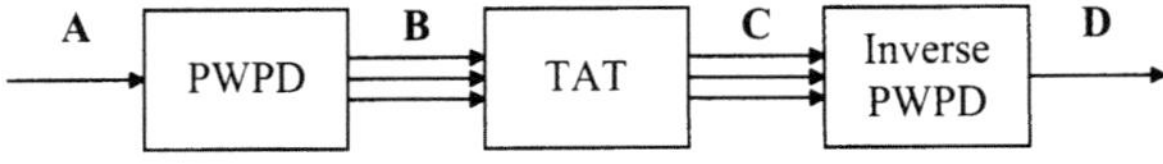

A: Noisy speech signal
B: Wavelet coefficients
C: Thresholded wavelet coefficients
D: Enhanced speech signal

Figure 1. System implementation for speech enhancement using PWPD and TAT.

speech signal to split its spectrum into critical banks corresponding to wavelet coefficients. The next stage is the TAT block, which capably suppresses the wavelet coefficients of noise and retains the wavelet coefficients of the speech signal in each critical bank. Finally, these thresholded wavelet coefficients are reconstructed to yield the enhanced speech signal via the inverse PWPD. Using speech signals corrupted by additive and real noises, experimental results demonstrate that the performance of the speech enhancement method proposed in this paper is better than those of conventional schemes. The remainder of this paper is organized as follows. Section 2 reviews the theories of the wavelet packet transform and the PWPD algorithm, respectively. Then, a detailed description of the TAT with TEO needed in the proposed speech enhancement method will be given in Section 3. Section 4 presents various experimental results of the proposed algorithm compared to the other methods. Finally, a conclusion will be given in Section 5.

2. Perceptual Wavelet Packet Decomposition Algorithm

This section will begin with a brief introduction to the theory of the wavelet packet transform. Then the description of the PWPD used in the proposed speech enhancement method is given. For more detailed mathematical discussions of the wavelet packet theory, see [23, 24].

2.1. Wavelet Packet Transform

The mathematical work of the wavelet packet transform was first proposed by Coifman [23]. Due to wavelet packets are generalized from the wavelet transform, the wavelet packet transform can also be implemented via filterbank structures. In fact, one of efficient ways to construct the discrete wavelet transform is to iterate a two-channel perfect reconstruction filterbank over the lowpass scaling function branch [23]. The approach given above was also called the Mallat algorithm [25]. However, this algorithm will result in a logarithmic frequency resolution which works well for most signals but not all. Moreover, the bandwidths of the subbands generated from the regular wavelet transform can not match those of the psychoacoustic model [18]. To overcome the above drawback, one can iterate the highpass wavelet branch of the Mallat algorithm tree as well as the lowpass scaling function branch. Such a wavelet decomposition produced by these arbitrary subband trees is called wavelet packet decomposition.

An illustrated example of a full binary tree for two-scale wavelet packet decomposition and reconstruction is given in Figure 2. In Figure 2, $h_0(n)$ and $h_1(n)$ are the analysis low-pass scaling filter and the high-pass wavelet filter, respectively, whereas $g_0(n)$ and $g_1(n)$ are the synthesis low-pass scaling filter and the high-pass wavelet filter, respectively. By Burrus et al. [23], for the perfect reconstruction property of the wavelet transform, these four filters have to related as

$$h_1(n) = (-1)^n g_0(1 - n), \quad g_1(n) = (-1)^n h_0(1 - n) \tag{3}$$

Also, the symbols $\downarrow 2$ and $\uparrow 2$ shown in Figure 2 denote the operation of downsampling by 2 and upsampling by 2, respectively. As shown in Figure 2, the wavelet packet transform can be constructed by the basic decomposition and reconstruction cells which are illustrated in Figure 3(a) and (b), respectively. In Figure 3(a), $\{c_0(n)\}_{n \in Z}$ denotes the input to the wavelet

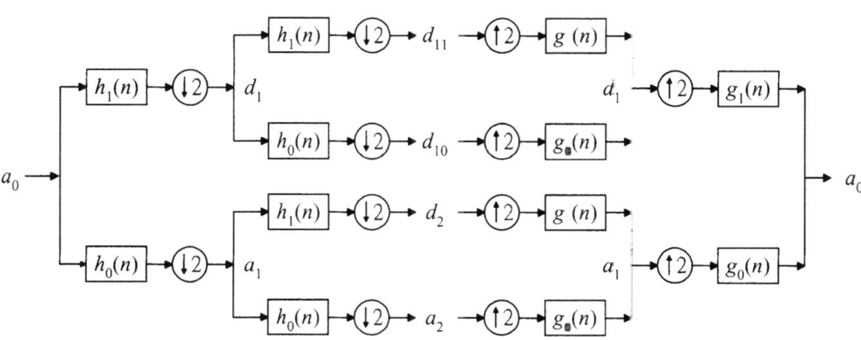

Figure 2.   The full binary tree for the two-scale wavelet packet transform.

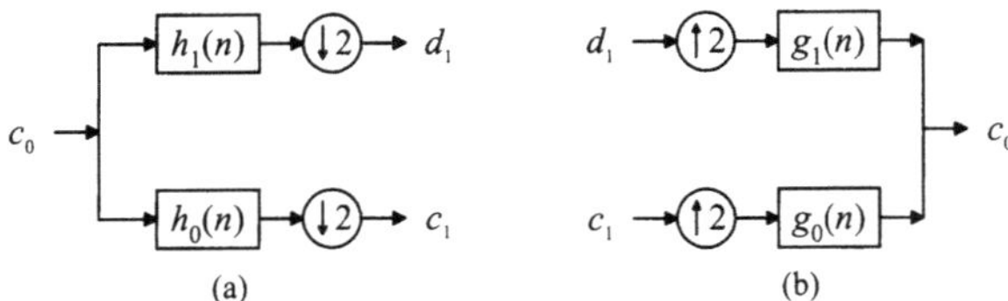

Figure 3. (a) The basic wavelet decomposition cell and (b) reconstruction cell.

decomposition cell and the outputs of this cell [23] are given by

$$c_1(k) = \sum_n h_0(n - 2k)c_0(n), \qquad (4)$$

$$d_1(k) = \sum_n h_1(n - 2k)c_0(n) \qquad (5)$$

where $c_1(k)$ and $d_1(k)$ are called the approximation coefficients and the detail coefficients of the first level wavelet decomposition of $c_0(n)$, respectively. And its corresponding wavelet reconstruction cell as shown in Figure 3(b) can be operated as

$$c_0(m) = \sum_k [g_0(2k - m)c_1(k) + g_1(2k - m)d_1(k)]. \qquad (6)$$

2.2. *Perceptual Wavelet Packet Decomposition*

As mentioned in the previous works [8, 9, 16–18], the PWPD algorithm is utilized to adjust the decomposition tree structure of the conventional wavelet packet transform in order to approximate the critical bands of the psychoacoustic model as close as possible. The primary reason for embedding the psychoacoustic model in the PWPD is that humans are capable of detecting the desired speech in a noisy environment without prior knowledge of the noise, see [26]. In the psychoacoustic model, frequency components of sounds can be integrated into critical bands that refer to bandwidths at which subjective response become significantly different [27]. Critical bands are of great importance in understanding many auditory phenomena, such as perception of loudness, pitch, and timbre. One class of critical band scales is called *Bark scale*. Based on the measurements by Zwicker et al. [28], the Bark scale z can be approximately expressed in terms of the linear frequency by

$$z(f) = 13 \arctan(7.6 \times 10^{-4} f)$$
$$+ 3.5 \arctan(1.33 \times 10^{-4} f)^2 \text{ [Bark]} \quad (7)$$

where f is the linear frequency in Hertz. The corresponding critical bandwidth (CBW) of the center frequencies can be expressed by

$$\text{CBW}(f_c) = 25 + 75(1 + 1.4 \times 10^{-6} f_c^2)^{0.69} \text{ [Hz]}$$
$$(8)$$

where f_c is the center frequency (unit: Hertz). Theoretically, the range of human auditory frequency spreads from 20 to 20000 Hz and covers approximately 25 Barks.

Due to the Bark scale is a function of linear frequency, the first step of constructing the PWPD is to set the sampling rate of speech signals in order to determine the valid Bark numbers. In this paper, the underlying sampling rate was chosen to be 8 kHz, yielding a bandwidth of 4 kHz. Within this bandwidth, there are approximately 17 critical bands as listed in Table 1 [27]. According to the specifications of center frequencies, CBW, lower and upper cutoff frequencies given in Table 1, the tree structure of the proposed PWPD can be constructed as shown in Figure 4(a). The corresponding frequency bandwidths of the PWPD tree is shown in Figure 4(b). It contains 16 decomposition cells with 5 decomposition stages to approximate these 17 critical bands. Hence, by the use of the PWPD, the input speech signal, namely $x(n)$, can be decomposed into 17 subbands which are corresponding to wavelet coefficient sets $w_{j,m}(k)$. More precisely, $w_{j,m}(k)$ defines the k-th coefficient of the m-th subband at j-th decomposition stage of PWPD, where $j = 3, 4, 5$, $m = 1, \ldots, 17$, and $k = 1, \ldots, N/2^j$. The resulting 17-band PWPD of the Bark scale and the CBW are also plotted in Figures 5 and 6, respectively.

3. TAT Algorithm Using TEO

Applying the TAT algorithm and the TEO to speech enhancement is first introduced by Bahoura and Rouat [10]. In [10], the TAT is cooperated with the conventional wavelet packet transform and the obtained results are closely similar to those from the Ephraim and Malah Filter (EMF) [4]. However, the TAT algorithm proposed in [10] still exists an over thresholding problem when the speech signal is just contaminated by slight noises. This comes from the fact that the frequency resolution provided by the 4-stage full wavelet packet transform is not appropriate to separate speeches from low-frequency noise, e.g. car noises. In addition, the algorithm [10] uses an absolute offset parameter to

Table 1.   The characteristics of critical bands under 4 kHz.

Critical band number	Center frequency (Hz)	CBW	Lower cutoff frequency (Hz)	Upper cutoff frequency (Hz)
1	50	–	–	100
2	150	100	100	200
3	250	100	200	300
4	350	100	300	400
5	450	110	400	510
6	570	120	510	630
7	700	140	630	770
8	840	150	770	920
9	1000	160	920	1080
10	1170	190	1080	1270
11	1370	210	1270	1480
12	1600	240	1480	1720
13	1850	280	1720	2000
14	2150	320	2000	2320
15	2500	380	2320	2700
16	2900	450	2700	3150
17	3400	550	3150	3700

distinguish speech frames from noise ones. From the original paper by Bahoura and Rouat, the aspect of the discrimination rule was only mentioned as an experimental finding. In the case of speech signal with small amplitude, the discrimination rule given in [10] will classify the speech frame into noise one and result in over thresholding.

Therefore, this paper modifies the original TAT algorithm [10] to overcome the above problems. The main modifications proposed in this paper are the uses of PWPD instead of conventional wavelet packet decomposition and using variance to distinguish speech frames from noise ones. Figure 7 is the flow chart of the proposed new TAT algorithm. Four subsystems called TEO, temporal masking construction, time adaptive threshold computation, and soft thresholding are contained in this TAT algorithm. The detailed descriptions of these four subsystems are described as follows:

3.1.   Teager Energy Operator

Similar to the TAT algorithm given in [10], the TEO is first applied to the wavelet coefficients to enhance the discriminability of speech and non-speech frames in each subband generated from PWPD. It has been shown [20–22] that the TEO is a powerful nonlinear operator

in various speech applications. The continuous form of the TEO was introduced by Kaiser [19, 20] and is given as

$$\Psi_c[y(t)] = \left(\frac{d}{dt}y(t)\right)^2 - y(t)\left(\frac{d^2}{dt^2}y(t)\right) \quad (9)$$

where $\Psi_c[\cdot]$ and $y(t)$ are the continuous TEO and a continuous speech signal, respectively. For a given bandlimited discrete speech signal $y(n)$, the discrete-time TEO can be approximated by

$$\Psi[y(n)] = y^2(n) - y(n+1)y(n-1). \quad (10)$$

The output of the TEO block called the TEO coefficients is given by

$$T_{j,m}(k) = \Psi[w_{j,m}(k)]. \quad (11)$$

where $w_{j,m}(k)$ are the wavelet coefficient sets generated from the PWPD.

The TEO applied to a digital signal in effect is looking at the signal as if a single sinusoid of amplitude A and frequency ω is passed through three adjacent points with index $n-1, n, n+1$; it yields as an output sequence the varying signal proportional to $A^2 \sin^2 \omega$ where the frequency ω is normalized with respect to the sampling frequency. This signal in essence is a measure

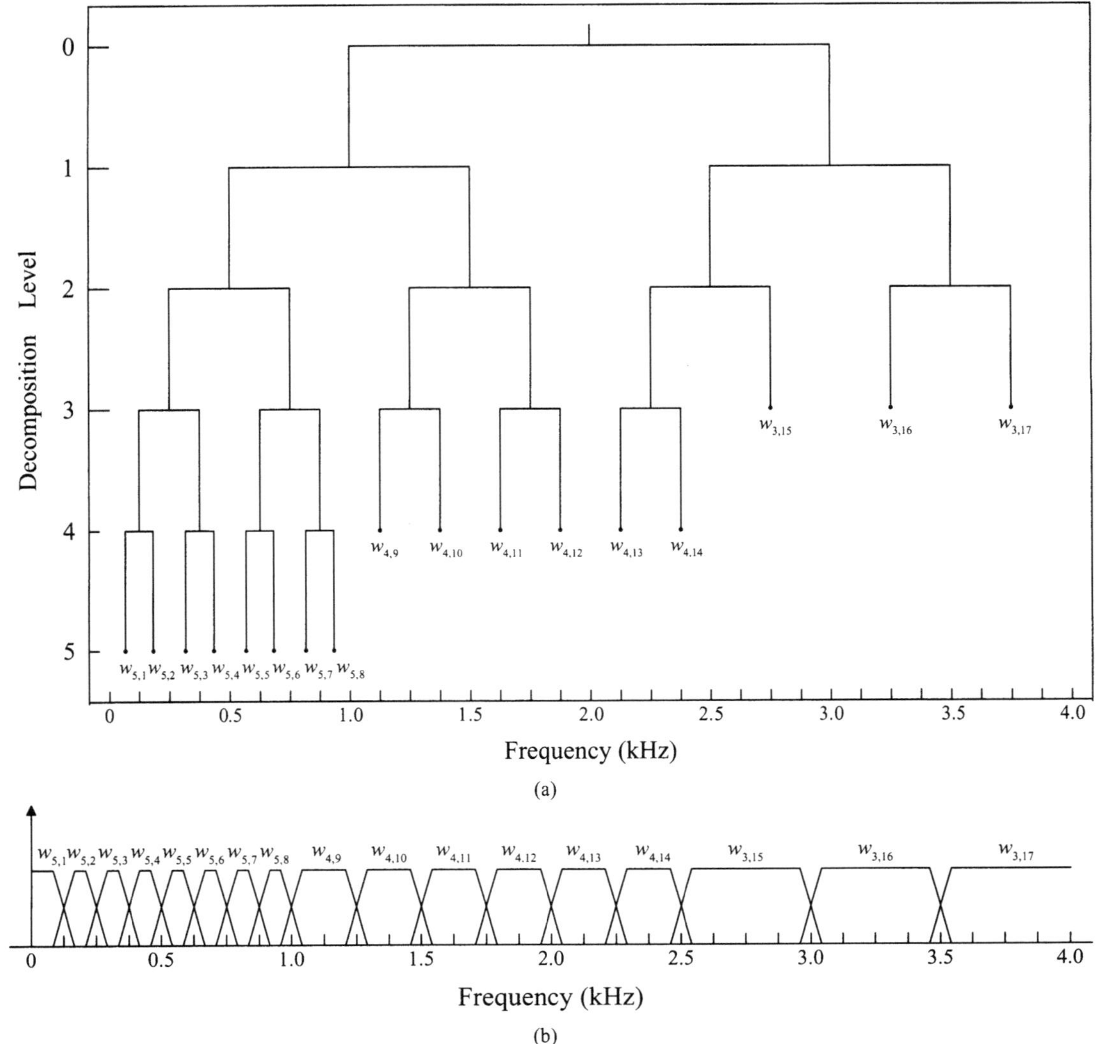

Figure 4. (a) The tree structure of the proposed PWPD. (b) The frequency bandwidths for the PWPD tree.

of the 'energy' in that signal as a function of time. It is thus important that the original digital signal consist primarily of a single component. In this application case, that of a single wavelet coefficient stream, it is primarily a single component signal and thus a valid application of the TEO.

3.2. Temporal Masking Construction

This paper applies a level-dependent FIR hamming window instead of an fixed second order IIR filter proposed in [10] to construct an initial mask $M_{j,m}(k)$. That is

$$M_{j,m}(k) = T_{j,m}(k) * H_j(k) \qquad (12)$$

where $*$ denotes the convolution operation and $H_j(k)$ is a hamming window of length $= 256/2^j$. The above operation is processed by sliding with overlapping half of window length. The main reason of using a level-dependent FIR filter is to retain the short-duration noise-like speech components, e.g. unvoiced and plosive consonants, in high-level PWPD.

3.3. Time Adaptive Threshold Computation

In this TAT algorithm, the threshold values are designed to match the temporal masking construction $M_{j,m}(k)$ obtained in the previous block. In other words, the threshold values should be adapted for speech segments and kept unchanged for non-speech ones. However, if

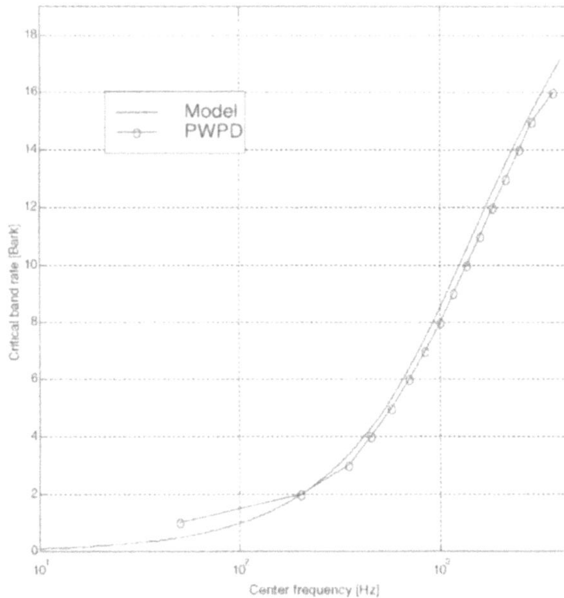

Figure 5.   Bark scale as a function of center frequency.

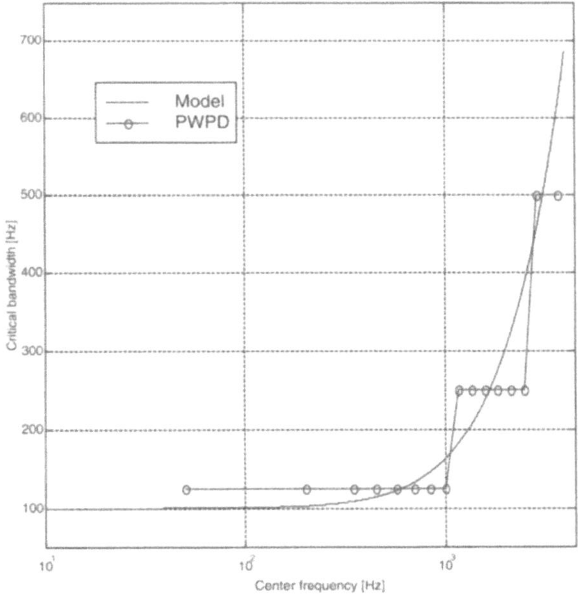

Figure 6.   Critical bandwidth as a function of center frequency.

the variance of $M_{j,m}(k)$ is equal to zero, then the corresponding subband will be regarded as a noise signal and its modulated temporal masking construction will set to be zero. That is,

$$M'_{j,m}(k) = 0. \tag{13}$$

Otherwise, the modulated temporal masking construction of each subband is obtained by

$$M'_{j,m}(k) = \left[\frac{M_{j,m}(k)}{\max(M_{j,m}(k))} \right]. \tag{14}$$

Note that the parameter of $M'_{j,m}(k)$ is close to 1 for speech segments and close to 0 for non-speech ones. Therefore, the time-adapted threshold values $\lambda_{j,m}(k)$ can be computed as

$$\lambda_{j,m}(k) = \lambda_j(1 - M'_{j,m}(k)) \tag{15}$$

where the level-dependent threshold λ_j [15] was defined by

$$\lambda_j = \sigma_j \sqrt{2 \log N}. \tag{16}$$

3.4.   Soft Thresholding

Finally, the proposed TAT algorithm is completed by the soft thresholding. That is,

$$\hat{w}_{j,m}(k) = \begin{cases} 0, & \text{if} |w_{j,m}(k)| < \lambda_{j,m}(k) \\ \text{sgn}(w_{j,m}(k))(|w_{j,m}(k) - \lambda_{j,m}(k)|), \\ & \text{otherwise} \end{cases} \tag{17}$$

where $\hat{w}_{j,m}(k)$ is the thresholded wavelet coefficients of the mth subband.

4.   Experimental Results

The proposed method was evaluated on natural speech signals corrupted by additive white Gaussian noises and real noisy speech signals. All of these tested speech signals are selected from the "Aurora 2" [29] database and are sampled at 8 kHz with 16-bit resolution of each sample. The software simulations were done using Matlab® 5.2 on a Pentium® II 500, Windows® 98 PC. The objective evaluation, i.e. SNR, and the subjective evaluation, i.e. mean opinion score (MOS), are applied to evaluate the performance of the speech enhancement method. The definition of the SNR is given by

$$\text{SNR} = 10 \log_{10} \left(\frac{\sum_{n=1}^{N} \{|s(n)|^2\}}{\sum_{n=1}^{N} \{|s(n) - \tilde{s}(n)|^2\}} \right) [\text{dB}] \tag{18}$$

where $s(n)$ and $\tilde{s}(n)$ denote the clean and enhanced speech signals, respectively. The MOS is performed by ten trained listeners under formal condition.

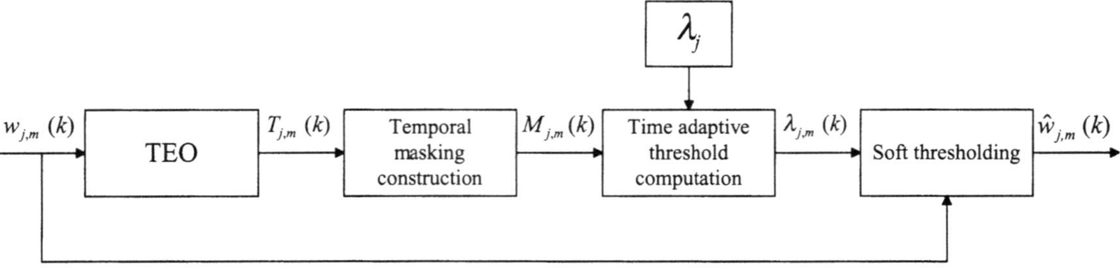

Figure 7. The flow chart of the proposed TAT algorithm.

4.1. On Choosing of Wavelet Filter

As many wavelet-based signal processing systems have shown [11], the choice of "mother wavelet" function or wavelet filter is important for the frequency selectively as well as the time domain resolution of the wavelet filterbank. In addition, the computational complexity of the wavelet filterbank is directly depended on the length of wavelet filter. Therefore, there exists a tradeoff in time-frequency resolution and computational complexity. In other words, the use of a longer wavelet filter will obtain a better frequency resolution but a heavy computational complexity.

As a result of the proposed speech enhancement method is based on the thresholding of the wavelet coefficients, some unmasked and aliased threshold noises may appear in sidelobes of the wavelet filters. Therefore, a sufficient stopband attenuation of the wavelet filter is required and the longer wavelet filters are needed. In this paper, the orthogonal wavelet filters including Daubechies, Coiflets, and Symlets are considered in the PWPD. In order to select an appropriate wavelet filter for the proposed speech enhancement method, a previous test experiment is given and its results are listed in Table 2. This test experiment was done using 100 test speech signals corrupted by different Gaussian white noises. The enhancement rate (*ER*) given in Table 2 is defined as

$$ER = \sum_{i=1}^{100} \{SNR[\tilde{s}_i(n)] - SNR[x_i(n)]\}/100 \quad (19)$$

where $SNR[\tilde{s}_i(n)]$ and $SNR[x_i(n)]$ denote the SNR of the i-th enhanced speech signal and the noisy speech signal, respectively. Considering the enhancement rate as well as the computational complexity given in Table 2, the Daubechies wavelet filter [24] with length of 10, which has the best *ER*/CPU time ratio, is recommended for the proposed method.

4.2. Performance of the Method for Additive Gaussian White Noise

The experiments executed in this subsection were performed using natural speeches corrupted by additive Gaussian white noises with different SNR's. Figures 8(a), (b), and (g) show the waveforms of clean, degraded, and enhanced speech signals, respectively. Furthermore, Figures 8(c)–(f) illustrate the corresponding waveforms of $w_{5,4}(k)$, $T_{5,4}(k)$, $\lambda_{5,4}(k)$, and $\hat{w}_{5,4}(k)$, respectively. The speech signal shown in Figure 8(a) corresponds to the sentence "Two-One-Two" spoken by a female speaker. In Figure 8, one observes that the noisy speech signal is contaminated by a Gaussian white noise with SNR of 5 dB, and the SNR of the

Table 2. The previous experimental results on choosing of wavelet filter.

	Daubechies					Coiflets		Symlets		
Wavelet filter type	D8	D10	D12	D14	D20	C12	C18	S12	S14	S16
Filter length	8	10	12	14	20	12	18	12	14	16
ER	3.2	3.6	3.6	3.7	3.9	3.5	3.7	3.6	3.6	3.7
CPU time (sec.)	0.35	0.39	0.60	0.94	1.78	0.59	1.18	0.60	0.89	0.97
ER/Time	9.14	9.23	6.00	'3.94	2.19	5.93	3.14	6.00	4.04	3.81

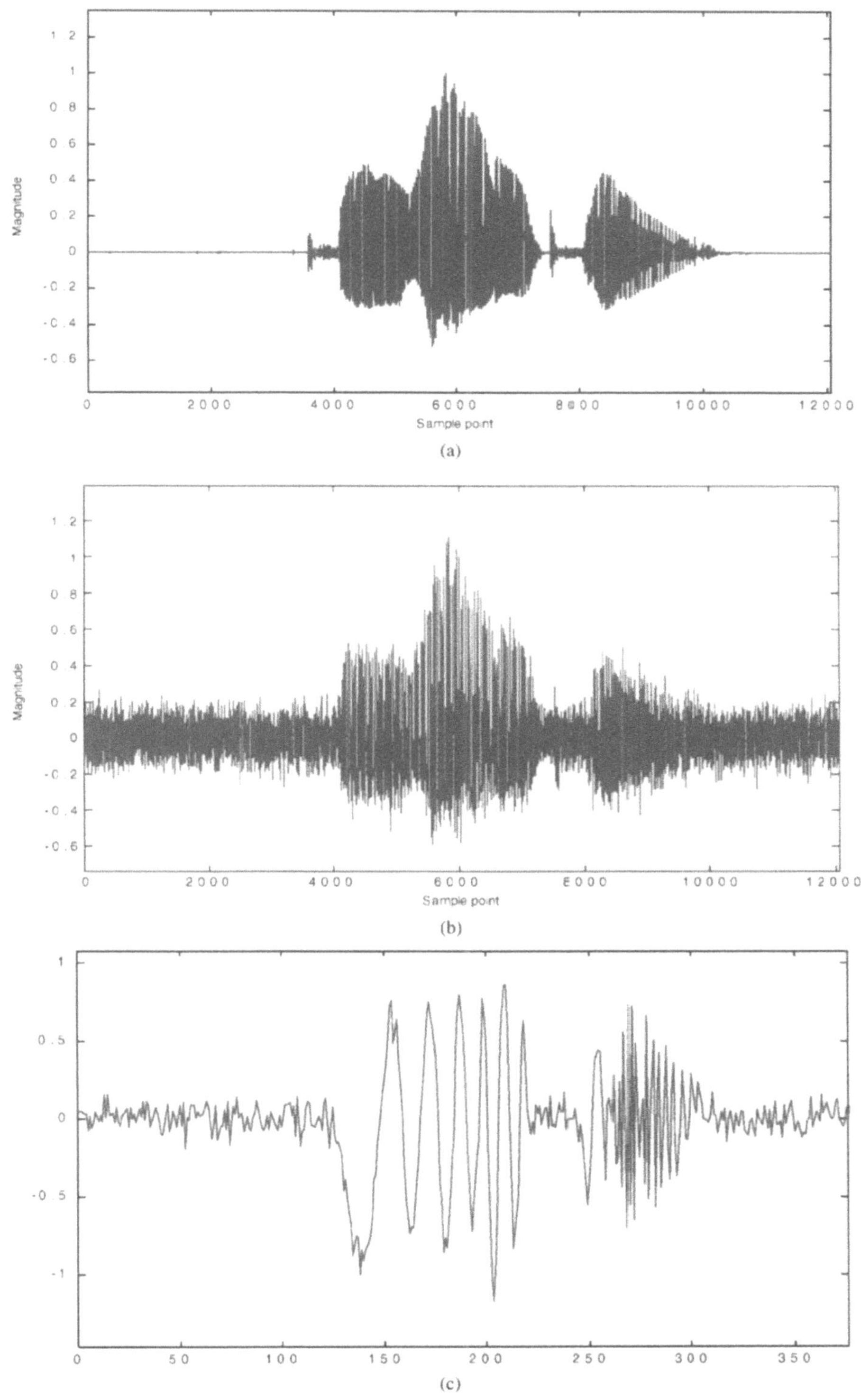

Figure 8. (a) Speech signal for the sentence "Two-One-Two". (b) Signal with an average SNR of 5 dB. (c) The corresponding waveform of $w_{5,4}(k)$. (d) The corresponding waveform of $T_{5,4}(k)$. (e) The corresponding waveform of $\lambda_{5,4}(k)$. (f) The corresponding waveform of $\hat{w}_{5,4}(k)$. (g) The enhanced speech signal with an average SNR of 9.6 dB. *(Continued on next page.)*

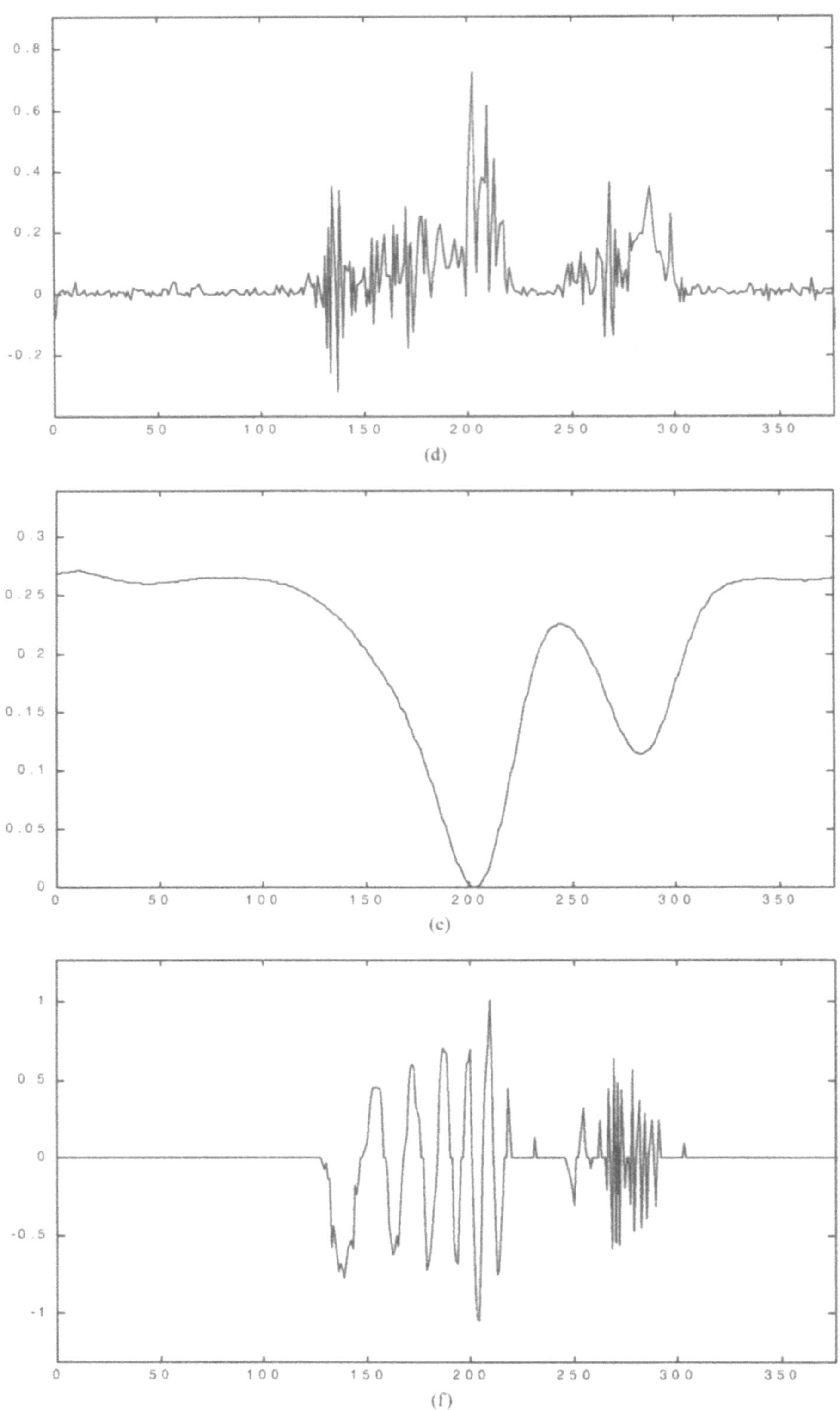

Figure 8.   (Continued).

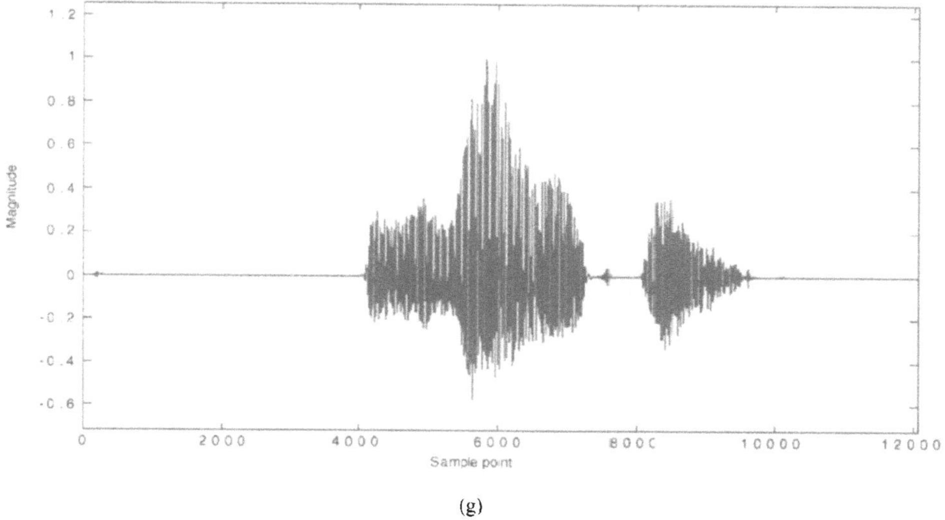

Figure 8. (*Continued*).

enhanced speech signal is improved to 9.6 dB. Also, it can be seen from the Figures 8(b) and (g) that the noise level is significantly attenuated, especially in the silence regions.

Finally, Table 3 demonstrates the performance of the proposed method in comparison with other wavelet-based and conventional EMF algorithms on speech signals corrupted by white Gaussian noise for SNR's ranging from -5 to 20 dB. It follows from Table 3 that the proposed method obtains the better performance than other speech enhancement methods. It is important to note that the over thresholding are avoided in the proposed method especially under the high SNR conditions.

Table 3. SNR tests for speech corrupted by additive Gaussian white noise.

Performance	Noise SNR					
	-5	0	5	10	15	20
Methods						
Proposed method	4.03	6.53	9.62	12.85	15.96	19.56
Method A	3.01	5.21	8.02	11.30	12.32	14.21
Method B	2.42	4.12	6.38	10.54	11.42	13.21
Method [4]	4.10	6.57	9.23	12.65	16.24	19.77
Method [10]	4.02	6.34	9.54	11.89	13.32	16.21

Method A: PWPD with a fixed threshold value.
Method B: Full 4-stage wavelet packet decomposition with a fixed threshold value.

4.3. Performance of the Method for Real Environment Noise

In this subsection, the proposed method was tested on speech signals recorded in real noisy environments including airport, car, exhibition, restaurant, and street. Figure 9(a) shows the noisy speech signal corresponding to the sentence "Nine-Nine-Seven-Five-Nine" and Figure 9(c) shows the speech waveform after using the proposed enhancement processing. The spectrograms of noisy and processed speech signals are illustrated in Figures 9(b) and (d), respectively. Using the proposed method, the improvement of the enhanced speech can be obtained and clearly seen in Figure 9. The MOS's of the speech signals in Figures 9(a) and (c) are 1.0 and 3.4, respectively. The MOS testing results on other noisy speech signals are also listed in Table 4. It is obvious that the proposed method obtains the best

Table 4. MOS tests for speech corrupted by real noise.

Noise type	Proposed method	Method A	Method B	Method [4]	Method [10]
Airport	3.7	3.2	2.9	3.4	3.6
Car	3.9	3.1	2.8	3.5	3.8
Exhibition	3.8	3.3	3.0	3.7	3.6
Restaurant	3.7	3.4	3.1	3.3	3.7
Street	3.5	3.2	2.8	3.5	3.5
Average	3.72	3.24	2.92	3.48	3.64

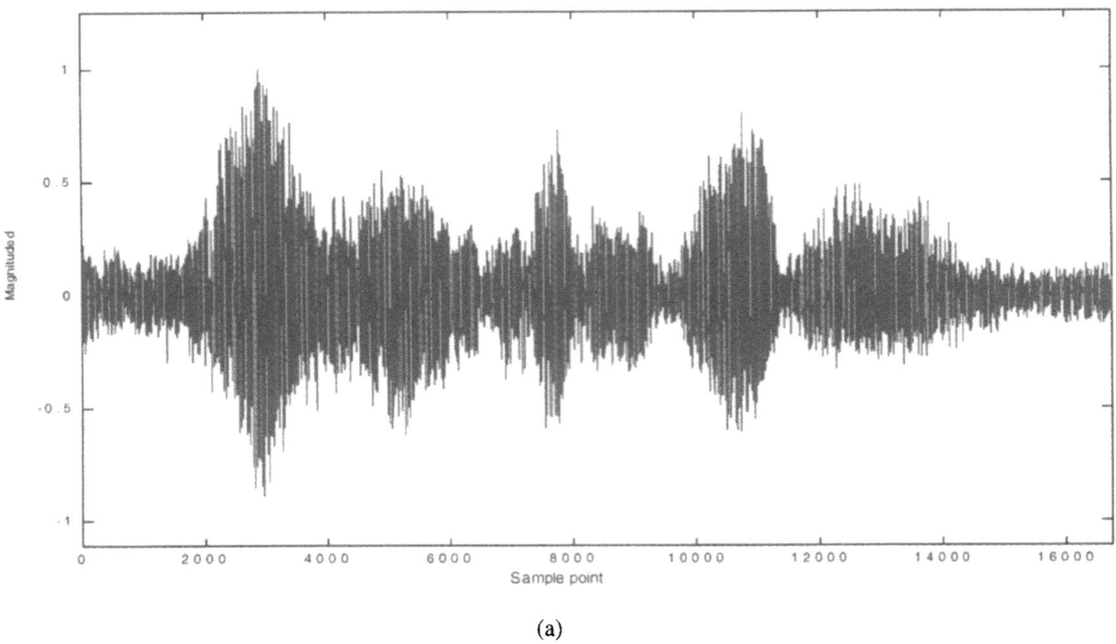

(a)

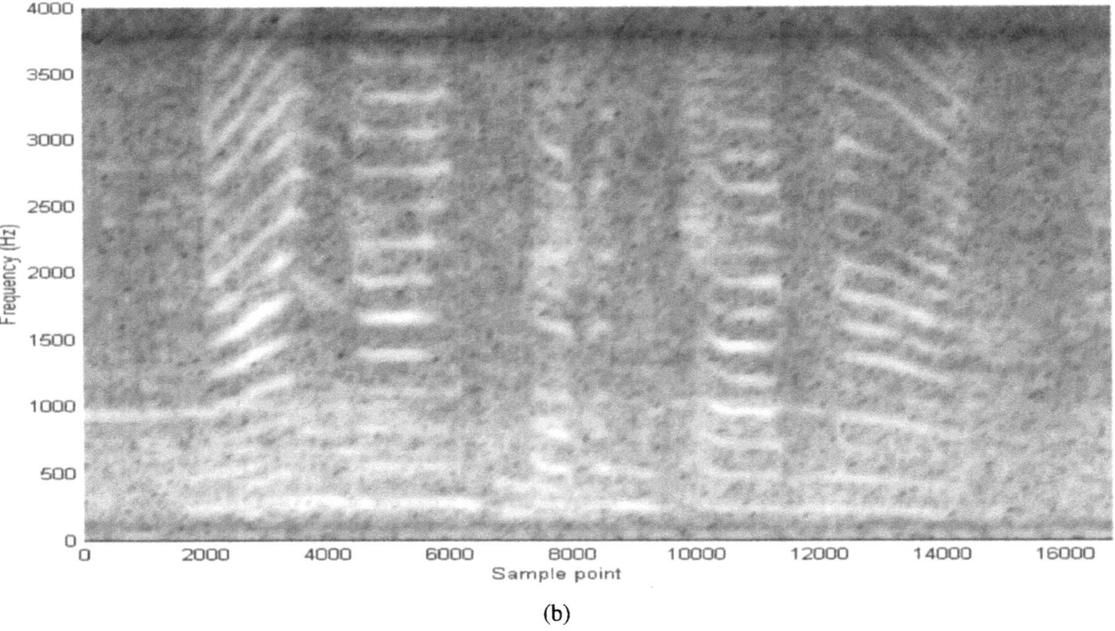

(b)

Figure 9.   Results of enhancement of speech degraded by airport noise. (a) Noisy speech. (b) Spectrogram of the noisy speech in (a). (c) Enhanced speech using the proposed method. (d) Spectrogram of the enhanced speech in (c).

(*Continued on next page.*)

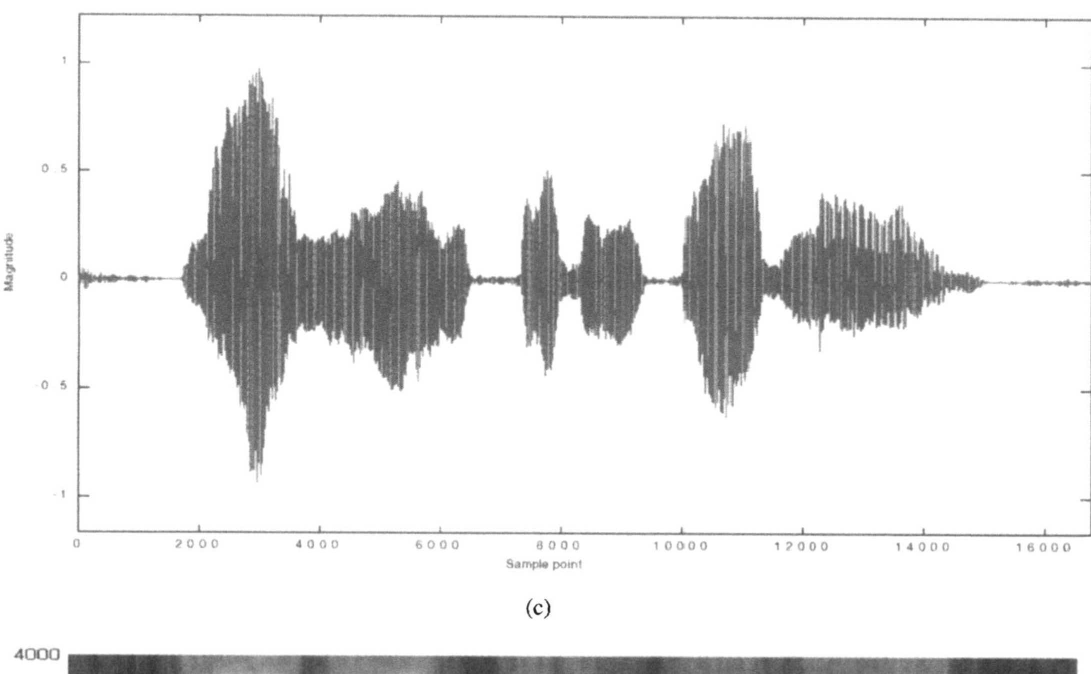

(c)

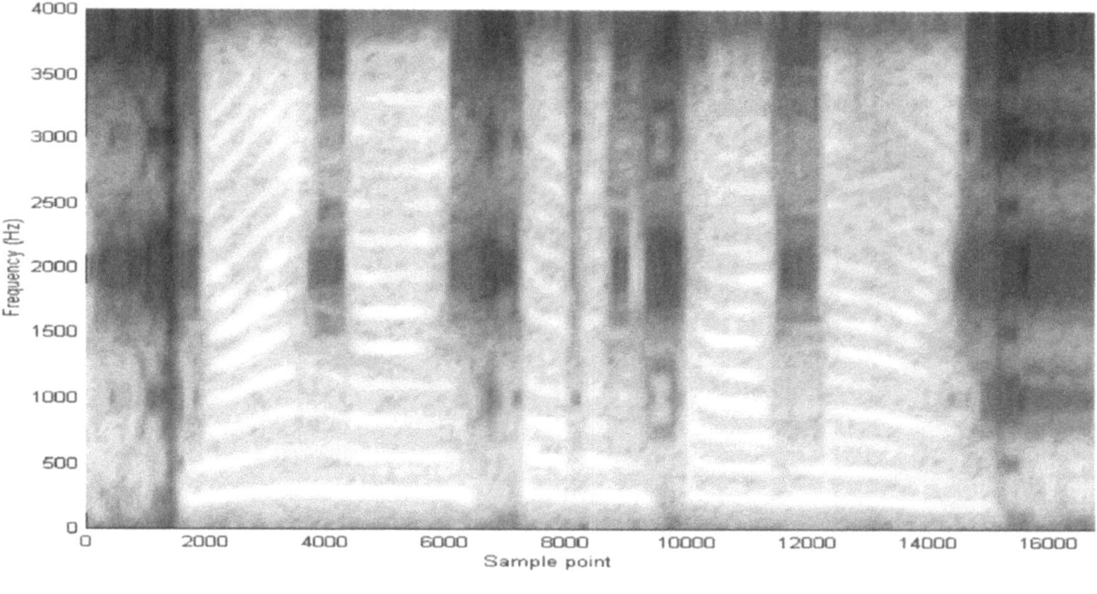

(d)

Figure 9.   (Continued).

perceptive enhanced qualities when compared with other speech enhancement methods.

5. Conclusions

An improved wavelet-based approach to the problem of speech enhancement using the PWPD and the TEO has been presented in this paper. By combining the advantage of the PWPD with that of the TEO, the proposed method avoids the over thresholding of speech segments especially when the speech is just corrupted by slight noises. The various experimental results show that this improved method has the better performance than other wavelet-based and conventional speech enhancement approaches. Furthermore, this improved method does not require a complicated estimation of the noise level or a prior knowledge of the noise SNR. In the future research works, the proposed speech enhancement method will be cooperated with speech recognition systems to increase their recognition rate under noisy environments.

Acknowledgment

This work was supported by the National Science Council (NSC), Taiwan, R.O.C., under Grant NSC90-2215-E-006-009.

References

1. B.H. Juang, "Recent Developments in Speech Recognition Under Adverse Conditions," in *Proceedings of Int. Conf. Spoken Language Process '90*, 1990, pp. 1113–1116.

2. J.H. Chen and A. Gersho, "Adaptive Postfiltering for Quality Enhancement of Coded Speech," *IEEE Trans. Speech and Audio Processing*, vol. 3, 1995, pp. 57–71.

3. R. Le Bouquin, "Enhancement of Noisy Speech Signals: Application To Mobile Radio Communications," *Speech Communication*, vol. 18, no. 1, 1996, pp. 3–19.

4. Y. Ephraim and D. Malah, "Speech Enhancement Using a Minimum Mean Square Error Short Time Spectral Amplitude Estimator," *IEEE Trans. Acoust. Speech Signal Processing ASSP-32*, 1984, pp. 1109–1121.

5. J. Meyer and K.U. Simmer, "Multi-Channel Speech Enhancement in a Car Environment Using Wiener Filtering and Spectral Subtraction," in *Proceedings of IEEE Int. Conf. Acoustics, Speech, and Signal Processing '97*, vol. 2, 1997, pp. 1167–1170.

6. B. Yegnanarayana, C. Avendano, H. Hermansky, and P. Satyanarayana Murthy, "Speech Enhancement Using Linear Prediction Residual," *Speech Communication*, vol. 28, 1999, pp. 25–42.

7. J.W. Seok and K.S. Bae, "Speech Enhancement with Reduction of Noise Components in The Wavelet Domain," in *Proceedings of IEEE Int. Conf. Acoustics, Speech, and Signal Processing '97*, vol. 2, 1997, pp. 1323–1326.

8. I. Pinter, "Perceptual Wavelet-Representation of Speech Signals and its Application to Speech Enhancement," *Computer Speech and Language*, vol. 10, no. 1, 1996, pp. 1–22.

9. B. Carneno and A. Drygajlo, "Perceptual Speech Coding and Enhancement Using Frame-Synchronized Fast Wavelet Packet Transform Algorithms," *IEEE Tans. Signal Processing*, vol. 47, no. 6, 1999, pp. 1622–1635.

10. M. Bahoura and J. Rouat, "Wavelet Speech Enhancement Based on the Teager Energy Operator," *IEEE Signal Processing Lett.*, vol. 8, 2001, pp. 10–12.

11. S.G. Chang, B. Yu, and M. Vetterli, "Adaptive Wavelet Thresholding for Image Denoising and Compression," *IEEE Trans. Image Processing*, vol. 9, 2000, pp. 1532–1546.

12. D.L. Donoho, "De-Noising by Soft-Thresholding," *IEEE Trans. Inform. Theory*, vol. 41, 1995, pp. 613–627.

13. D.L. Donoho and I.M. Johnstone, "Ideal Spatial Adaptation by Wavelet Shrinkage," *Biometrika*, vol. 81, 1994, pp. 425–455.

14. D.L. Donoho, "Unconditional Bases are Optimal Bases for Data Compression and Statistical Estimation," *Applied and Computational Harmonic Analysis*, vol. 1, 1994, pp. 100–115.

15. I.M. Johnstone and B.W. Silverman, "Wavelet Threshold Estimators for Data with Correlated Noise," *J. Roy. Statist. Soc. B*, vol. 59, 1997, pp. 319–351.

16. O. Farooq and S. Datta, "Mel Filter-Like Admissible Wavelet Packet Structure for Speech Recognition," *IEEE Signal Processing Letters*, vol. 8, no 7, 2001, pp. 196–198.

17. R. Sarikaya, B. Pellom, and J.H.L. Hansen, "Wavelet Packet Transform Features with Application to Speaker Identification," in *NORSIG-98, IEEE Nordic Signal Processing Symposium*, Vigso, Denmark, 1998, pp. 81–84.

18. P. Srinivasan and L.H. Jamieson, "High Quality Audio Compression Using an Adaptive Wavelet Decomposition and Psychoacoustic Modeling," *IEEE Trans. Signal Processing*, vol. 46, no. 4, 1998, pp. 1085–1093.

19. J.F. Kaiser, "On a Simple Algorithm to Calculate the "Energy" of a Signal," in *Proceedings of IEEE Int. Conf. Acoustics, Speech, and Signal Processing '90*, 1990, pp. 381–384.

20. J.F. Kaiser, "Some Useful Properties of Teager's Energy Operator," in *Proceedings of IEEE Int. Conf. Acoustics, Speech, and Signal Processing '93*, 1993, pp. 149–152.

21. F. Jabloun, A.E. Cetin, and E. Erzin, "Teager Energy Based Feature Parameters for Speech Recognition in Car Noise," *IEEE Signal Processing Lett.*, vol. 6, 1999, pp. 259–261.

22. G. Zhou, J.H.L. Hansen, and J.F. Kaiser, "Nonlinear Feature Based Classification of Speech Under Stress," *IEEE Trans. Speech and Audio Processing*, vol. 9, 2001, pp. 201–216.

23. C.S. Burrus, R.A. Gopinath, and H. Guo, *Introduction to Wavelets and Wavelet Transforms, A Primer*, Upper Saddle River, Nj: Prentice-Hall, 1998.

24. I. Daubechies, *Ten Lectures on Wavelets*, CBMS, SIAM Publ., 1992.

25. S. Mallat, "Multifrequency Channel Decomposition of Images and Wavelet Model," *IEEE Trans. Acoustic, Speech and Signal Processing*, vol. 37, 1989, pp. 2091–2110.

26. O. Ghitza, "Auditory Model and Human Performance in Tasks Related to Speech Coding and Speech Recognition," *IEEE Trans. Speech and Audio Processing*, vol. 2, 1994, pp. 115–132.

27. L. Rabiner and B.H. Juang, *Fundamental of Speech Recognition*, Upper Saddle River, NJ: Prentice-Hall, 1993.

28. E. Zwicker and E. Terhardt, "Analytical Expressions for Critical-Band Rate and Critical Bandwidth as a Function of Frequency," *JASA*, vol. 68, 1980, pp. 1523–1525.

29. See http://www.icp.inpg.fr/ELRA/aurora2.html.

Shi-Huang Chen was born in Tainan, Taiwan, R.O.C. in 1972. He received the B.S. and M.S. degrees in electrical engineering from the Kaohsiung Polytechnic Institute, Kaohsiung, Taiwan, in 1994

and 1996, respectively, and the Ph.D. degree in electrical engineering from the National Cheng Kung University, Tainan, Taiwan, in 2002. From 2001 to 2003 he was a Research Engineer at the AVXing Inc., Kaohsiung, Taiwan. He is now an Assistant Professor at the Department of Computer Science and Information Engineering, Shu-Te University, Kaohsiung, Taiwan. His research interests are in the areas of wavelet transforms, image, video, speech, and audio coding technologies, and multimedia communication standards.
shchen@cad.ee.ncku.edu.tw

Jhing-Fa Wang is now a professor in National Cheng Kung University, Tainan, Taiwan. He received his Master and Bachelor degrees in the department of Electrical Engineering from National Cheng Kung University in 1979 and 1973, respectively and Ph.D. degree in the Department of Computer Science and Electrical Engineering from Stevens Institute of Technology, U.S.A. in 1983. He is now a board Chairman of Journal of The Chinese Institute of Electrical Engineering. He is now also on the Board of Governors of IEEE Taipei Section. He got outstanding awards from Institute of Information Industry in 1991 and National Science Council in 1990, 1995, and 1997, respectively. His current research areas include VLSI/CAD, speech recognition, speech coding, optical character recognition, and natural language processing. He has developed a Mandarin speech recognition system called Venus-Dictate known as a pioneering system in Taiwan. He has published about 81 journal papers and 194 conference papers since 1983. He was elected as an IEEE Fellow in 1999 for contributions to software-hardware co-development of large-vocabulary Mandarin speech processing and recognition systems.
wangjf@server2.iie.ncku.edu.tw

Journal of VLSI Signal Processing 36, 141–151, 2004

Use of Microphone Array and Model Adaptation for Hands-Free Speech Acquisition and Recognition

JEN-TZUNG CHIEN AND JAIN-RAY LAI
*Department of Computer Science and Information Engineering, National Cheng Kung University,
Tainan, 70101 Taiwan, Republic of China*

Received October 30, 2001; Revised June 5, 2002; Accepted September 3, 2002

Abstract. This paper presents a combined microphone array and model adaptation algorithm for hands-free speech recognition. Our purpose is to remove the inconvenience of using head-mounted/hand-holding microphone in conventional speech recognizer. To improve the speech quality with car noise interference, a linear microphone array is applied and acted as robust acquisition system. A time-domain coherence measure (TDCM) is applied to reliably estimate the time delay for speech signals collected by different microphones. The estimated delay is adopted in a delay-and-sum beamformer for speech enhancement. Further, we adapt the speech hidden Markov models to get close to the acoustic conditions of the enhanced test speech for robust speech recognition. In acquisition and recognition experiments using connected Chinese digits, we found that TDCM can effectively estimate the time delay. The increase in the speech sampling rate is helpful to determine the time delay. Incorporating the model adaptation scheme significantly reduces the recognition errors with moderate computation overhead.

Keywords: microphone array, delay-and-sum beamformer, coherence measure, model adaptation, speech enhancement, speech recognition

1. Introduction

In realistic environments, the room reverberation and the noises from fans, babble, music, air conditioners, etc, deteriorate the speech signals a great deal. Especially, in the environments for teleconferencing, car driving and mobile communication, it is convenient and efficient to use the far-talking hands-free microphone in voice-driven human-machine interface. However, the signal-to-noise ratio (SNR) of recorded speech is relatively low which creates unacceptable speech recognition performance. In general, two directions are available to resolve the problems occurring in hands-free speech recognition. First, the robust acquisition system in a microphone array can depress room echoes/noises and increase the speech SNR in the signal space. Desirable speech recognition performance can therefore be obtained. Although the speech signals are enhanced, mismatches still exist between the enhanced speech and

the speech hidden Markov models (HMM's) trained from a large data pool containing variabilities in speakers, microphones and noises. Consequently, alternative methods aimed at adapting the original HMM's to fit the test speech must be used. After adaptation in model space, the speech recognition robustness can be assured. We endeavored to improve the hands-free speech recognition in signal space as well as model space.

Array signal processing [1] has been employed in speech technology for a long time. In the literature, the optimal spacing and gain of linear microphone arrays were analyzed for speech data acquisition [2]. The microphone array was also applied for location estimation [3–6], target tracking [7], noise reduction [8] and echo cancellation [9, 10]. Because of its beamforming power, the microphone array has been recognized as an important vehicle for speech enhancement. Beamforming schemes with directivity pattern optimization

[11, 12] and multiresolution wavelet transform combinations [13, 14] were effective in increasing the SNR of degraded speech. Microphone array applications have been widely extended to hands-free distant speech recognition in cars and reverberation rooms [15–18]. A 3-D Viterbi algorithm with the search space composed of talker direction, input frames and HMM states was developed for speech recognition [19]. The huge search space however made this algorithm impractical to use. It is quite crucial to incorporate a model adaptation technique in microphone array based speech recognition. An incremental maximum *a posteriori* (MAP) training/adaptation algorithm was performed after beamforming signal enhancement [20]. Other adaptation methods like MAP adaptation [21] and maximum likelihood linear regression (MLLR) [22] were also successful in improving the recognition robustness of beamforming speech [23, 24].

In this study, a delay-and-sum beamformer was adopted to increase the SNR of distant speech contaminated with car noise. To alleviate the mismatch between the beamforming speech and trained HMM's, instantaneous MLLR adaptation was performed to obtain desirable recognition results. Generally, a delay-and-sum beamformer is designed to enhance noisy speech via the *time delay compensation* of different microphones. The time delay estimation becomes critical for beamforming schemes. Omologo and Svaizer [3] calculated the time delay by finding the maximum coherence measure using the crosspower-spectrum phase (CSP). The CSP measure was also used to localize multiple sound sources [6]. Further, Yamada et al. [4] presented an algorithm called speaker localization using an arrayed microphone (SLAM) for time delay compensation. They detected the most likely speaker direction (or equivalently time delay) possessing the largest spatial power spectrum. Both CSP and SLAM are operated in the frequency domain. In this study, we applied a time-domain coherence measure (TDCM) for efficient and effective time delay estimation in presence of car noise. To improve the delay estimation resolution, we also increased the speech signal sampling rate. In the connected Chinese digit recognition experiments, the combined mechanism with delay-and-sum beamforming using TDCM and model adaptation using MLLR achieved the best results among the various methods. In the next section, a system overview of the proposed speech acquisition and recognition will be given. The combined technique using TDCM and MLLR is described in Section 3. In Section 4, a se-

ries of SNR, recognition rate and recognition speed experiments for evaluating the TDCM and combined TDCM and MLLR performance is described. Finally, a set of conclusion drawn from this study is provided in Section 5.

2. System Overview

The microphone array based speech acquisition and recognition flow diagram is plotted in Fig. 1. The microphone array served as the input device for recording the distant speech data. The microphone time delay was then estimated and forwarded to the delay-and-sum beamformer for speech enhancement. The SLAM and TDCM algorithms were investigated for the time delay estimation. The mismatch between the enhanced speech and the speaker-independent (SI) HMM's trained from near microphone utterances were overcome using the instantaneous MLLR model adaptation. The enhanced test speech was recognized using the adapted HMM's.

In the delay-and-sum beamformer, it is assumed that a speech plane wave comes from the direction θ to an equally spaced array composed of M microphones. Let d denote the distance between the adjacent microphones and x_t^i represent the speech signal from microphone i at time t. When microphone i receives the plane wave at time t, the wave will proceed with distance R in sound velocity $C = 342$ m/s and arrive at microphone $i + 1$ with time delay

$$\tau = \frac{R}{C} = \frac{d \cos \theta}{C}. \tag{1}$$

Figure 2 illustrates the time delay from the adjacent microphones. In theory, if the sensitivity of the microphone hardware is identical, the speech signal from microphone i the relation to those from microphone

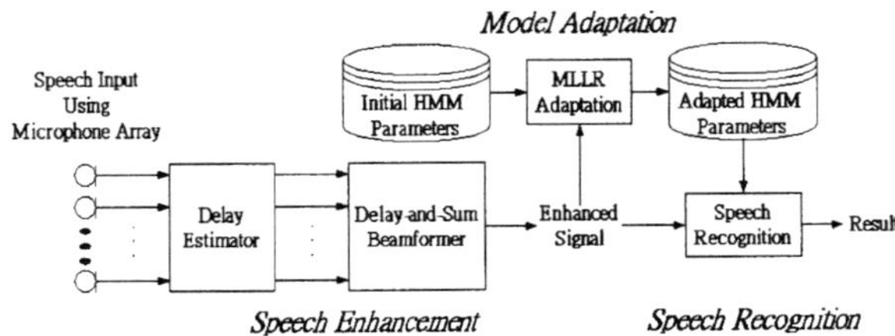

Figure 1. System overview of microphone array based speech acquisition and recognition.

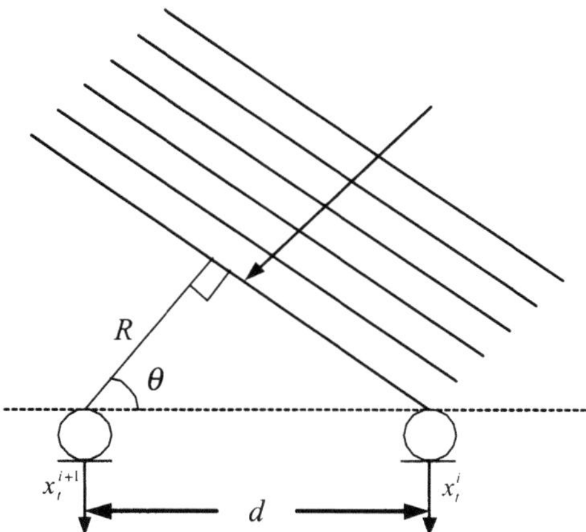

Figure 2.   Illustration for time delay of adjacent microphones.

$i + 1$ and the first microphone

$$x_t^i = x_{t+\tau}^{i+1} = x_{t-(i-1)\tau}^1.\tag{2}$$

The delay-and-sum beamforming principle is to enhance the signal quality via time delay compensation. Namely, at time t, the delay-and-sum beamformer sums the compensated signals from various microphones and produces

$$\hat{x}_t = \sum_{i=1}^{M} x_{t+(i-1)\tau}^i = \sum_{i=1}^{M} x_t^i \exp\left\{ j2\pi f(i-1)\frac{d\cos\theta}{C} \right\},\tag{3}$$

where $j = \sqrt{-1}$ and the array signals $\{x_t^i, i = 1, \ldots, M\}$ are assumed to be a sequence of complex sinusoidal signals in frequency f despite the fact that the speech signals are real. It turns out that the delay estimation accuracy deeply affects the quality of the enhanced signal.

Previously, the SLAM algorithm [4] was used for the delay-and-sum beamformer. The time delay is detected indirectly using three steps. First, the speech frames of M microphones $\{x_t^i\}$ at time t are transformed into K-point frequency-domain signals $\{X_t^i(k)\}$ via the fast Fourier transform (FFT). Integer k denotes the harmonic index. In the second step, the *spatial power spectrum* corresponding to each sound wave direction angle $1° \leq \theta \leq 180°$ is calculated over M microphones and K harmonics as follows

$$P_t(\theta) = \sum_{k=0}^{K-1} \left| \sum_{i=1}^{M} X_t^i(k) \exp\left\{ j2\pi f_k(i-1)\frac{d\cos\theta}{C} \right\} \right|^2.\tag{4}$$

Each harmonic k corresponds to the frequency f_k. In the final step, the speaker direction is detected by searching the most likely angle $\hat{\theta}$ owing to the maximum spatial power spectrum

$$\hat{\theta} = \arg\max_{\theta} P_t(\theta).\tag{5}$$

The search for the most likely angle $\hat{\theta}$ is actually equivalent to the time delay estimation as shown in (1). One may select a time moment t with reliable discrete points for a frame $\{x_i^{\cdot}\}$ to calculate $P_t(\theta)$. The optimal angle $\hat{\theta}$ is then detected and shared for speech signal delay compensation at all time moments t. One may also accumulate the spatial power spectra $P_t(\theta)$ over a period of time segment. The speaker direction $\hat{\theta}$ is then identified with improved accuracy.

3.   Combined Technique Using TDCM and MLLR

A combined technique using the TDCM in the delay-and-sum beamformer for speech enhancement and the instantaneous MLLR HMM adaptation for robust speech recognition is presented.

3.1.   TDCM

The proposed algorithm determines the time delay by *directly* calculating the *time-domain coherence measure* (TDCM) of the speech signals at various distant microphones. Let the speech frames of microphone i and microphone $i + 1$ starting at time moment t be represented by $x_t^i, x_{t+1}^i, \ldots, x_{t+N}^i$ and $x_{t+\tau}^{i+1}, x_{t+1+\tau}^{i+1}, \ldots, x_{t+N+\tau}^{i+1}$, respectively. The time delay of the neighboring microphones is expressed with τ discrete points. Each frame contains $N + 1$ points. In theory, the time delay τ could be estimated such that the coherence function

$$c_t^{i,t+1}(\tau) = \sum_{n=0}^{N} x_{t+n}^i x_{t+n+\tau}^{i+1},\tag{6}$$

between the frames of the neighboring microphones i and $i + 1$ has the largest value. More generally, we may define the TDCM of an M-microphone array at time moment t with time delay τ by *summing the coherent functions of all possible microphone pairs* (i, p)

$$C_t(\tau) = \sum_{i=1}^{M-1} \sum_{p>i}^{M} c_t^{i,p}((p-i)\tau). \tag{7}$$

The speech plane wave arrives at the succeeding microphone p with $p - i$ times the time delay τ later than the preceding microphone i. This measure can be viewed as a crucial factor to reflect the *degree of cross correlation* in the speech signals in a microphone array. Accordingly, the time delay estimation involves finding the delay that generates the maximum TDCM

$$\hat{\tau} = \arg\max_{\tau} C_t(\tau). \tag{8}$$

Similar to the SLAM algorithm, we could estimate the most likely time delay using either the TDCM of a time moment t or the accumulated TDCM covering a segment/sentence, i.e.

$$\tilde{\tau} = \arg\max_{\tau} \sum_{t} C_t(\tau). \tag{9}$$

Johnson and Dudgeon [1] addressed and employed the inter-signal correlation between the target signal and the reflected signal to detect the presence of the return signal as well as estimate the time delay. The correlation within an observation interval was derived via the hypothesis testing principle. A cross-correlation scheme was exploited to determine the time delay and the speaker's location in a noise-free room [25]. The cross-correlation was computed over the windowed signals of two microphones. In this study, the TDCM was calculated throughout the whole sentence and using the set of all possible microphone pairs $\{(1, 2), (1, 3), \ldots, (1, M), (2, 3), \ldots, (2, M), \ldots, (M - 1, M)\}$. Incorporating more information from different microphone pairs is helpful to achieve better time delay. The estimated time delay was forwarded to the delay-and-sum beamformer for speech enhancement in the presence of various car noises. In these experiments, two delay estimation cases were investigated using (8) and (9) and compared in terms of the SNR, recognition rate and computation time.

3.2. Instantaneous MLLR Adaptation

Although the delay-and-sum beamformer is helpful for enhancing distant speech, the mismatch between the enhanced test speech and the trained HMM's still exists. This mismatch is especially serious in the case of numerous training speakers and training utterances spoken near the microphone. To resolve the mismatch problem, the existing HMM's are adjusted to match the acoustics of the enhanced speech. Speech recognition could be improved using the adapted HMM's. Leggetter and Woodland [22] proposed a maximum likelihood linear regression (MLLR) approach to speaker adaptation. This approach was intended to estimate the cluster-dependent transformation functions to adapt/transform the overall HMM mean vectors to a new speaker using very limited adaptation data. The transformation of HMM mean vector μ_{nm} of state n and mixture component m can be written using an affine function

$$\hat{\mu}_{nm} = \mathbf{A}_c \mu_{nm} + \mathbf{b}_c, \tag{10}$$

where μ_{nm} is attributed to the transformation cluster c containing a scaling matrix $\mathbf{A}_c$ and a bias vector $\mathbf{b}_c$. The instantaneous/self MLLR adaptation was used because only the feature vectors of the enhanced test speech $\hat{\mathbf{X}} = \{\hat{\mathbf{x}}_t\}$ are available for model adaptation. The most likely regression parameters are estimated via maximizing likelihood function

$$(\hat{\mathbf{A}}_c, \hat{\mathbf{b}}_c) = \arg\max_{(\mathbf{A}_c, \mathbf{b}_c)} P(\hat{\mathbf{X}} \mid \mathbf{A}_c, \mathbf{b}_c, \lambda = \{\mu_{nm}, \Sigma_{nm}\}), \tag{11}$$

where the HMM parameters λ contain mean vectors $\{\mu_{nm}\}$ and covariance matrices $\{\Sigma_{nm}\}$. Because the estimation in (11) is an incomplete data problem, the expectation-maximization (EM) algorithm [26] is applied to merge the missing HMM state and mixture component labels and derive the formulation for the regression parameters $(\hat{\mathbf{A}}_c, \hat{\mathbf{b}}_c)$. Readers may refer to [22] for the details. MLLR adaptation was adopted in this study instead of MAP adaptation [21] because transformation-based MLLR adaptation can adapt the overall HMM mean vectors via cluster-dependent regression parameters. However, MAP adaptation adjusts only the HMM units appearing in the adaptation data. In the case of very limited test data $\hat{\mathbf{X}}$, the MLLR adaptation is a preferable choice. Moreover, the instantaneous MLLR adaptation is unsupervised such that two recognition passes are required [27]. The first pass is devoted to obtain the test utterance transcription. Given the transcription, the regression parameters are estimated and used for modeling the adaptation. The test utterance $\hat{\mathbf{X}}$

is finally recognized using the adapted HMM's through the second recognition pass.

4. Experiments

4.1. *Speech Databases and Baseline System*

We conducted a series of speech acquisition and recognition experiments on connected Chinese digits to examine the combined TDCM and MLLR algorithm. Two databases were collected for evaluation. The first database was recorded in an office environment via four close-talking microphones. There were 1400 utterances of connected digits from 70 males and 70 females. This database was used to train the SI HMM parameters. A second database was prepared for testing by using a multi-channel SigC31-4 board, produced by Signalogic, Inc. Readers may refer http://www.signalogic.com/sigc31.htm for detail board specifications. Four equally spaced omnidirectional condenser microphones (type ECM9D) were connected to a SigC31-4 board. The configuration for speaker, noise and microphone array is displayed in Fig. 3. It was set up at one side of a rectangular room with 5-meter length, 4-meter width and 3-meter height. The reverberation time of this room was about 0.35 sec. In general, we used this configuration to evaluate the time delay estimation methods. The realistic speech recognition in moving car environment could not be well simulated using the point noise source. Herein, all test utterances were distant with a distance of 1.2 meters and uttered by 12 males and 3 females. The speaker direction relative to the microphone array had

an angle of 60 degrees. The background noise signals from a medium car (YULON SENTRA 1.6) driving at the speeds of 0 km/h (standby condition), 50 km/h (downtown condition) and 90 km/h (freeway condition) were sampled. These noise samples were played through a loudspeaker at an angle of 105 degrees to serve as the noise source. Each test speakers separately uttered 30 sentences for the three driving conditions. The connected digits in the training and test sentences were randomly spoken and had an averaged length of six digits. The speech was digitized at 16-bit accuracy at an 8 kHz sampling rate. The proposed algorithm was investigated using the SNR in decibel (dB) and the recognition performance in word error rate (WER). The SNR's and WER's were averaged over the utterances of 15 test speakers. For each noisy speech utterance x_t, the noise signal segment n_t was automatically detected via HMM based silence detector. The SNR was calculated by

$$\text{SNR(dB)} = 10 \log_{10} \frac{\sigma_x^2 - \sigma_n^2}{\sigma_n^2} \qquad (12)$$

where σ_x^2 and σ_n^2 were the corresponding sample variances for x_t and n_t respectively. The clean speech and noise signals were assumed to be independent. The baseline condition was set for the case where no delay-and-sum beamformer and model adaptation was used. The SNR's and WER's of four microphones were averaged to serve as the baseline results.

In the speech recognition experiments, each Chinese digit was modeled using a left-to-right seven-state continuous-density HMM without state skipping. There were 73 HMM states (70 for ten digits, 1 for pre-silence, 1 for post-silence and 1 for silence within connected digits). Each HMM state was composed of four mixture components. A speech frame with 256 discrete points was characterized by a feature vector with 12 Mel-frequency cepstral coefficients, 12 delta cepstral coefficients, one delta log energy and one delta delta log energy. The spatial power spectrum in (4) was implemented with the resolution of θ at 5 degrees. The covariance matrix Σ_{nm} in the HMM and scaling matrix $\mathbf{A}_c$ in MLLR were simplified into the diagonal for all experiments. Only one regression class was adopted in MLLR because the adaptation data (i.e. one test sentence) was too limited. The SNR's for the four individual microphone array channels under various noise samples are listed in Table 1. The SNR's between the four channels are different because the

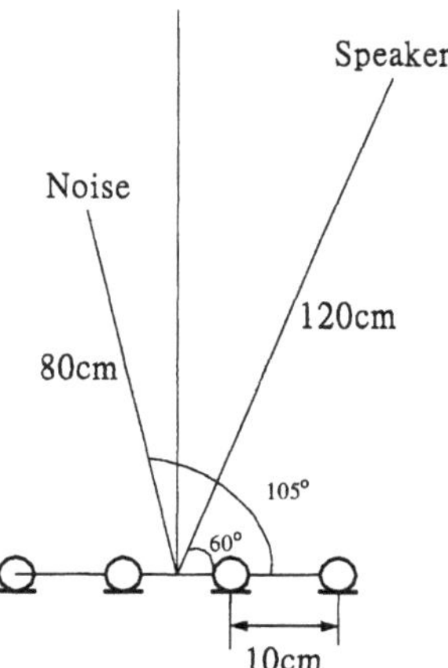

Figure 3.   Configuration for speaker, noise and microphone array.

Table 1. Signal-to-noise ratios (dB) for four microphone array channels under various noise conditions. The signal-to-noise ratios averaged over four channels are the baseline results.

Channel/Noise condition	90 km/h	50 km/h	0 km/h
Channel 1	3.89	5.35	7.67
Channel 2	4.53	6.43	9.08
Channel 3	2.78	4.36	6.14
Channel 4	3.2	5.06	7.1
Average (baseline)	3.6	5.3	7.5

Table 2. Word error rates (%) for four microphone array channels under various noise conditions. The word error rates averaged over four channels are the baseline results.

Channel/Noise condition	90 km/h	50 km/h	0 km/h
Channel 1	55.1	52.1	47
Channel 2	53.4	48.3	42
Channel 3	58.7	54.8	51
Channel 4	57.2	53	46.5
Average (baseline)	56.1	52.1	46.6

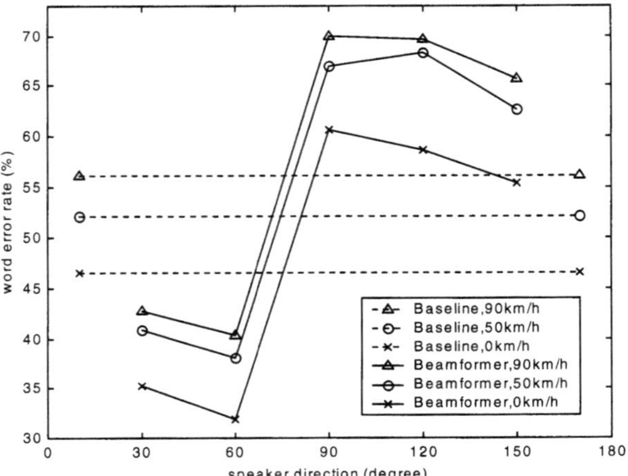

Figure 4. Comparison of word error rates (%) for the baseline system and delay-and-sum beamformer when the speaker direction is assumed to be 30°, 60°, 90°, 120° and 150° relative to the microphone array.

microphone hardware sensitivity has inherent variations. The second channel has the highest SNR and the third the lowest. On average, the baseline SNR's were 3.6, 5.3 and 7.5 dB for distant speech in presence of 90 km/h, 50 km/h and 0 km/h noise conditions, respectively. The SNR differences were small because the noise source was distant as well. Using the data acquired from the four channels, WER's without model adaptation were carried out and are listed in Table 2. The differences between the four microphones are consistent in WER's and SNR's. The baseline WER's were 56.1%, 52.1% and 46.6% under noise conditions of 90 km/h, 50 km/h and 0 km/h, respectively.

4.2. *Effect of Speaker Direction on the Delay-and-Sum Beamformer*

We were primarily interested in examining the time delay compensation effect on the delay-and-sum beamformer. To conduct this examination, we determined the microphone array system speaker location from a pre-specified direction. Once the direction was determined, the associated time delay was calculated using (1). In Fig. 4, the WER's of the baseline system and delay-and-sum beamformer are compared when the speaker direction is assumed to be 30°, 60°, 90°, 120° and 150° relative to the microphone array. We can see that the WER's of delay-and-sum beamformer were even worse than the baseline WER's in the 90°, 120° and 150° cases. When the speaker direction was specified at 60°, the WER's were greatly reduced to 40.4%, 38.1% and 31.9% under 90 km/h, 50 km/h and 0 km/h noise conditions, respectively. In fact, this direction is an exact match with the configuration shown in Fig. 3. From these results, we learned that speaker direction determination is crucial to estimate the desirable time delay. The resulting delay-and-sum beamformer could be significantly better than the baseline system for WER's.

4.3. *Comparison of the Baseline System, SLAM and TDCM Algorithms*

Subsequently, we investigated the time delay compensation performance for the delay-and-sum beamformer using the previous SLAM [4] and the proposed TDCM algorithms. The speech frame of an utterance with the highest energy was heuristically selected to calculate the spatial power spectrum in (4) and the TDCM in (7) even though the selected frame might be flawed in the case of very noisy signals. Generally, the microphone array beam-form is frequency dependent and determined by the spacing of the microphone elements. In our array configuration, the equally spaced linear microphone array produces the beam-form mainly in

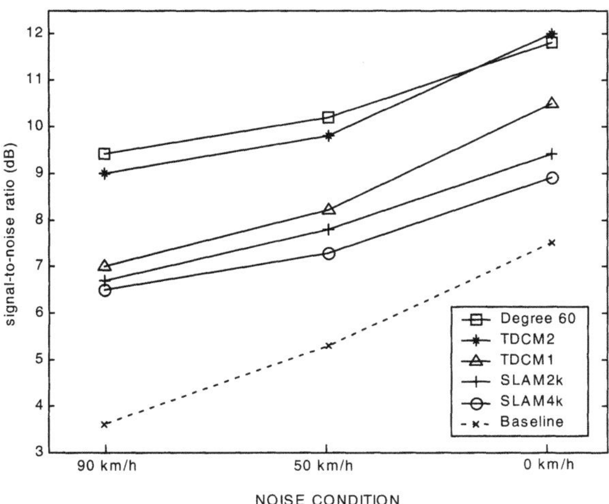

Figure 5. Comparison of signal-to-noise ratios (dB) for the baseline system, SLAM4k, SLAM2k, TDCM1, TDCM2 and delay-and-sum beamformer with the exact speaker direction at 60° under various noise conditions.

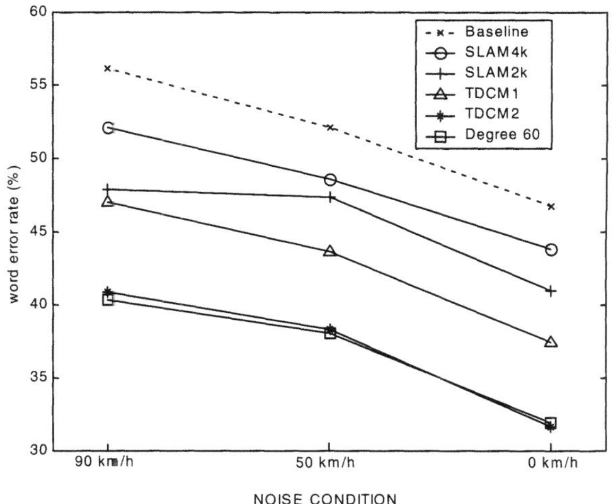

Figure 6. Comparison of word error rates (%) for the baseline system, SLAM4k, SLAM2k, TDCM1, TDCM2 and delay-and-sum beamformer with the exact speaker direction at 60° under various noise conditions.

the frequency range 500–2000 Hz. We examined two SLAM variants in which the frequency range for calculating the spatial power spectrum in (4) was varied. SLAM2k and SLAM4k are designed to cover the frequency bands 500–2000 Hz and 0–4000 Hz, respectively. TDCM is a time-domain method that is not affected by the frequency coverage. For comparison, we also examined TDCM variants using a single speech frame in (8) (denoted by TDCM1) and using all frames of a test utterance in (9) (denoted by TDCM2). SLAM4k, SLAM2k, TDCM1 and TDCM2 were employed in a delay-and-sum beamformer for speech acquisition. The output speech signals could be evaluated using the SNR measure. In Fig. 5, the SNR's of the baseline system and delay-and-sum beamformer using SLAM4k, SLAM2k, TDCM1, TDCM2 and the exact speaker direction at 60° are illustrated. We found that various delay-and-sum beamformer cases do increase the SNR's. A beamformer with the exact speaker direction obtains the highest SNR's that can be viewed as upper the bounds of the delay-and-sum beamformer. The proposed TDCM2 could attain SNR's close to the upper bounds. Figure 5 also indicates that TDCM is superior to SLAM for various variants. SLAM2k attains higher SNR's than SLAM4k. This is because SLAM4k adopts the spatial power spectrum covering the frequency range above 2000 Hz, which might cause the spatial sampling theorem to go unsatisfied. Spatial aliasing occurs and the resulting spatial power spectrum is underestimated. Under 50 km/h noise condi-

tions, the SNR's increased from 5.3 dB in the baseline to 7.3 dB in the SLAM4k, 7.8 dB in the SLAM2k, 8.2 dB in the TDCM1 and 9.8 dB in the TDCM2. The SNR of the TDCM2 approached 10.2 dB using the exact speaker direction.

Another set of experiments was executed to assess the recognition performance of the enhanced speech signals produced using various delay-and-sum beamformer cases. The WER results are shown in Fig. 6. SLAM and TDCM outperformed the baseline delay-and-sum beamformer. The TDCM performance was better than that of SLAM. It is promising that the TDCM2 WER's were comparable to those from exact time delay compensation. For example, the baseline WER 46.7% was reduced to 43.8%, 41%, 37.5%, 31.7% and 31.9% using SLAM4k, SLAM2k, TDCM1, TDCM2 and exact time delay compensation, respectively, under 0 km/h noise conditions. In our experimental setup, the true time delay value due to the speaker location is 1.23 discrete points calculated using $8000 \cdot (10 \cos 60°/32400)$. We can determine the mean and variance for the estimated time delay errors in the test utterances with respect to the true value. As listed in Table 3, the TDCM consistently achieved a smaller estimation error than SLAM2k. From these results, we are confident that TDCM could effectively estimate the time delay for the delay-and-sum beamformer. In the following experiments involving the SLAM2k variant are discussed.

Table 3. Mean and variance of the delay estimation error using SLAM and TDCM.

	SLAM	TDCM
Mean	0.297	0.152
Variance	0.018	0.011

4.4. Sampling Rate Effect on the SLAM and TDCM Algorithms

In the analysis of time delays estimated using SLAM and TDCM, we learned that most test utterances had delays less than three discrete points. The delay points were calculated at sampling rate of 8 kHz. To reinforce the time delay resolution, we up-sampled the acquired speech prior to using the SLAM and TDCM algorithms. Once the delay point is determined, the enhanced speech using delay-and-sum beamformer is obtained and down-sampled to 8 kHz for speech recognition. SLAM and TDCM were employed at 16 kHz sampling rate in comparison to those without the up-sampling process. The WER's of the baseline system, SLAM and TDCM algorithms at sampling rates of 8 kHz and 16 kHz are compared in Fig. 7. The TDCM1 was performed. We can see that the increase in sampling rate did reduce the WER's using SLAM and TDCM. The WER reduction produced by SLAM and TDCM was significant. The SLAM algorithm could be improved because it indi-

rectly searches the time delay discrete point via optimal estimation of speaker direction $\hat{\theta}$. The delay point is determined by $(\text{SamplingRate}) \cdot d \cos \hat{\theta}/C$. Increasing the sampling rate could raise the precision of the estimated delay point. Under 90 km/h noise conditions, the WER's 47.9% of SLAM and 47% of TDCM at 8 kHz sampling rate were decreased to 43.9% of SLAM and 43.1% of TDCM at 16 kHz sampling rate. However, this improvement was achieved at the expense of extra computation for changing the sampling rate and processing double discrete points. We continued the following experiments using a sampling rate at 8 kHz.

4.5. Recognition Results and Speeds Using Instantaneous MLLR Adaptation

We have doubts about the fitness of the static HMM's for distant speech recognition. The instantaneous MLLR adaptation therefore serves as a post-processor to estimate the adaptive HMM's for robust speech recognition. As listed in Table 4, the WER's of baseline plus MLLR was averaged over the MLLR adaptation in four microphone channels. The WER's are plotted in Fig. 8 when the delay-and-sum beamformers using SLAM, TDCM1 and TDCM2 were combined with the MLLR adaptation. These results reveal that the performance is greatly improved by performing additional MLLR adaptation. In case of the 0 km/h noise conditions, the baseline WER of 46.6% dropped to 28.6% using the baseline plus MLLR. This is similar to 29.9% for SLAM plus MLLR. However, when TDCM1 and TDCM2 plus MLLR were performed, the WER's were considerably reduced to 25.2% and 21.1%, respectively. We compare the WER's of TDCM2 with and without instantaneous MLLR adaptation in Table 5. Moreover, the recognition speeds measured in

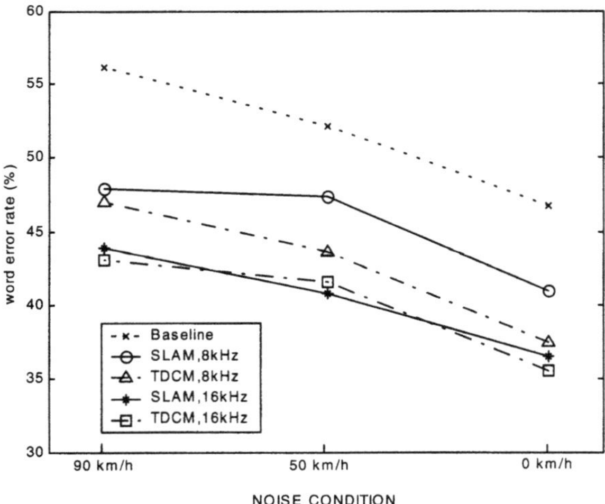

Figure 7. Comparison of word error rates (%) for the baseline system, SLAM and TDCM algorithms at sampling rates of 8 kHz and 16 kHz.

Table 4. Word error rates (%) for four microphone array channels using instantaneous MLLR adaptation under various noise conditions. The word error rates averaged over four channels are the baseline plus MLLR results.

Method/Noise condition	90 km/h	50 km/h	0 km/h
Channel 1 + MLLR	33.4	31.7	29
Channel 2 + MLLR	33.1	29.9	25.5
Channel 3 + MLLR	36.2	33.4	31.6
Channel 4 + MLLR	34.5	31.4	28.3
Baseline + MLLR	34.3	31.6	28.6

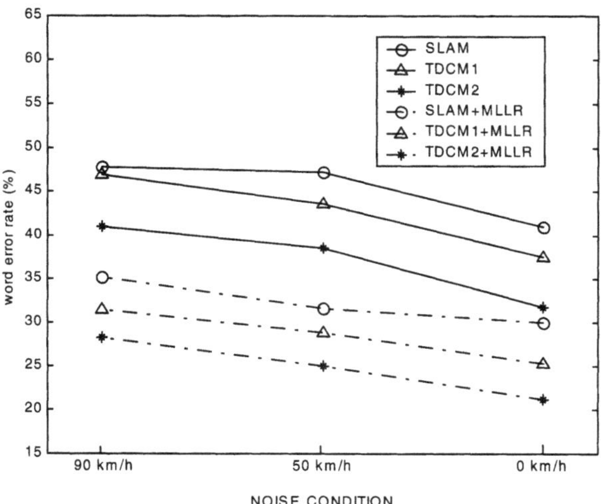

Figure 8. Comparison of the word error rates (%) for the SLAM, TDCM1 and TDCM2 with and without instantaneous MLLR adaptation under various noise conditions.

seconds per utterance are reported in Table 6. The baseline system, SLAM, TDCM1 and TDCM2 with and without MLLR were simulated on a PENTIUM II 350 personal computer with 128 Mega bytes RAM. We can see that SLAM used nearly double the computation cost compared to the baseline system. Using time-domain scheme TDCM1, the cost becomes smaller. The computation cost of TDCM2 however, was close to that of SLAM because the coherence measures of all frames were computed in TDCM2. We found that instantaneous MLLR adaptation requires extra 0.22 seconds, which is moderate for most applications.

Table 5. Word error rates (%) for TDCM2 with and without instantaneous MLLR adaptation.

Method/Noise condition	90 km/h	50 km/h	0 km/h
Without MLLR	40.9	38.4	31.7
With MLLR	28.2	24.9	21.1

Table 6. Computation costs (second per utterance) for the baseline system, SLAM, TDCM1 and TDCM2 with and without instantaneous MLLR adaptation.

	Baseline	SLAM	TDCM1	TDCM2
Without MLLR	0.28	0.58	0.41	0.56
With MLLR	0.50	0.79	0.63	0.77

5. Conclusion

We presented a microphone array based speech acquisition and recognition framework. The delay-and-sum beamforming technique was adopted to increase the SNR of a distant speech signal. To improve the conventional delay-and-sum beamformer, a time-domain coherence measure was applied where the time delay can be effectively detected as if the beamformer knew the exact speaker direction. The instantaneous MLLR adaptation was appended to the TDCM based delay-and-sum beamformer such that the mismatch between the distant test speech and the trained HMM's could be partially compensated to achieve robust speech recognition. After a series of experiments on the acquisition and recognition of Chinese connected digits, we summarize our observations as follows.

1. The speaker direction substantially affects the performance of the delay-and-sum beamformer. The exact direction is beneficial for improving the SNR and WER in a microphone array system.
2. The proposed TDCM is efficient and superior to the previous SLAM algorithm in terms of SNR and WER. TDCM could attain performance comparable to the delay-and-sum beamformer with the exact speaker direction.
3. Time delay estimation using SLAM and TDCM could be processed at a higher sampling rate. Smaller WER's could be obtained.
4. Instantaneous MLLR adaptation as a post processor for the delay-and-sum beamformer does reduce the WER's greatly at the limited expense of computation cost.

All of these observations were confirmed using the microphone array speech data in the presence of car noise samples. In the future, we need to further investigate the proposed methods in realistic car environments.

Acknowledgments

The authors acknowledge anonymous reviewers providing valuable comments, which considerably improved the quality of this paper. This work was partially supported by the National Science Council, Taiwan, ROC, under contract NSC89-2213-E-006-151.

References

1. D.H. Johnson and D.E. Dudgeon, *Array Signal Processing: Concepts and Techniques*, Prentice-Hall, Inc., 1993.
2. H.F. Silverman, "Some Analysis of Microphone Arrays for Speech Data Acquisition," *IEEE Transactions on Acoustics, Speech, and Signal Processing*, vol. ASSP-35, no. 12, 1987, pp. 1699–1711.
3. M. Omologo and P. Svaizer, "Acoustic Event Localization Using a Crosspower-Spectrum Phone Phase Based Technique," *IEEE Proceedings of International Conference on Acoustics, Speech, and Signal Processing (ICASSP)*, vol. 2, 1994, pp. 273–276.
4. T. Yamada, S. Nakamura, and K. Shikano, "Robust Speech Recognition with Speaker Localization by a Microphone Array," in *Proceedings of International Conference on Spoken Language Processing (ICSLP)*, 1996, pp. 1317–1320.
5. M.S. Brandstein, J.E. Adcock, and H.F. Silverman, "A Closed-Form Location Estimator for Use with Room Environment Microphone Arrays," *IEEE Transactions on Speech and Audio Processing*, vol. 5, no. 1, 1997, pp. 45–50.
6. T. Nishiura, T. Yamada, S. Nakamura, and K. Shikano, "Localization of Multiple Sound Sources Based on a CSP Analysis with a Microphone Array," in *IEEE Proceedings of International Conference on Acoustics, Speech, and Signal Processing (ICASSP)*," vol. 2, 2000, pp. 1053–1056.
7. Y. Nagata and H. Tsuboi, "A Two-Channel Adaptive Microphone Array with Target Tracking," in *Proceedings of European Conference on Speech Communication and Technology (EUROSPEECH)*, 1997, pp. 343–346.
8. M. Mizumachi and M. Akagi, "Noise Reduction by Paired-Microphones Using Spectral Subtraction," in *IEEE Proceedings of International Conference on Acoustics, Speech, and Signal Processing (ICASSP)*, vol. 2, 1998, pp. 1001–1004.
9. Q.-G. Liu, B. Champagne, and P. Kabal, "A Microphone Array Processing Technique for Speech Enhancement in a Reverberant Space," *Speech Communication*, vol. 18, 1996, pp. 317–334.
10. M. Dahl and I. Claesson, "Acoustic Noise and Echo Canceling with Microphone Array," *IEEE Transactions on Vehicular Technology*, vol. 48, no. 5, 1999, pp. 1518–1526.
11. M. Dorbecker, "Small Microphone Arrays with Optimized Directivity for Speech Enhancement," in *Proceedings of European Conference on Speech Communication and Technology (EUROSPEECH)*, 1997, pp. 327–330.
12. H. Saruwatari, S. Kajita, K. Takeda, and F. Itakura, "Speech Enhancement Using Nonlinear Microphone Array with Complementary Beamforming," in *IEEE Proceedings of International Conference on Acoustics, Speech, and Signal Processing (ICASSP)*, vol. 1, 1999, pp. 69–72.
13. D. Mahmoudi, "A Microphone Array for Speech Enhancement Using Multiresolution Wavelet Transform," in *Proceedings of European Conference on Speech Communication and Technology (EUROSPEECH)*, 1997, pp. 339–342.
14. D. Mahmoudi and A. Drygajlo, "Combined Wiener and Coherence Filtering in Wavelet Domain for Microphone Array Speech Enhancement," in *IEEE Proceedings of International Conference on Acoustics, Speech, and Signal Processing (ICASSP)*, vol. 1, 1998, pp. 385–388.
15. M. Inoue, S. Nakamura, T. Yamada, and K. Shikano, "Microphone Array Design Measures for Hands-Free Speech Recognition," in *Proceedings of European Conference on Speech Communication and Technology (EUROSPEECH)*, 1997, pp. 331–334.
16. D. Giuliani, M. Matassoni, M. Omologo, and P. Svaizer, "Use of Different Microphone Array Configurations for Hands-Free Speech Recognition in Noisy and Reverberant Environment," in *Proceedings of European Conference on Speech Communication and Technology (EUROSPEECH)*, 1997, pp. 347–350.
17. R. Aubauer, R. Kern, and D. Leckschat, "Optimized Second Order Gradient Microphone for Hands-Free Speech Recordings in Cars," in *Proceeding of Workshop on Robust Methods for Speech Recognition in Adverse Conditions*, 1999, pp. 191–194.
18. J. Bitzer, K.U. Simmer, and K.D. Kammeyer, "Multi-Microphone Noise Reduction Techniques for Hands-Free Speech Recognition—A Comparative Study," in *Proceeding of Workshop on Robust Methods for Speech Recognition in Adverse Conditions*, 1999, pp. 171–174.
19. T. Yamada, S. Nakamura, and K. Shikano, "Hands-Free Speech Recognition Based on 3-D Viterbi Search Using a Microphone Array," in *IEEE Proceedings of International Conference on Acoustics, Speech, and Signal Processing (ICASSP)*, vol. 1, 1998, pp. 245–248.
20. J.E. Adcock, Y. Gotoh, D.J. Mashao, and H.F. Silverman, "Microphone-Array Speech Recognition via Incremental MAP Training," in *IEEE Proceedings of International Conference on Acoustics, Speech, and Signal Processing (ICASSP)*, vol. 2, 1996, pp. 897–900.
21. J.L. Gauvain and C.-H. Lee, "Maximum *a posteriori* Estimation for Multivariate Gaussian Mixture Observations of Markov Chains," *IEEE Transactions on Speech and Audio Processing*, vol. 2, 1994, pp. 291–298.
22. C.J. Leggetter and P.C. Woodland, "Maximum Likelihood Linear Regression for Speaker Adaptation of Continuous Density Hidden Markov Models," *Computer Speech and Language*, vol. 9, 1995, pp. 171–185.
23. D. Giuliani, M. Matassoni, M. Omologo, and P. Svaizer, "Experiments of HMM Adaptation for Hands-Free Connected Digit Recognition," in *IEEE Proceedings of International Conference on Acoustics, Speech, and Signal Processing (ICASSP)*, vol. 1, 1998, pp. 473–476.
24. J. Kleban and Y. Gong, "HMM Adaptation and Microphone Array Processing for Distant Speech Recognition," in *IEEE Proceedings of International Conference on Acoustics, Speech, and Signal Processing (ICASSP)*, vol. 3, 2000, pp. 1411–1414.
25. H.F. Silverman and S.E. Kirtman, "A Two-Atage Algorithm for Determining Talker Location from Linear Microphone Array Data," *Computer Speech and Language*, vol. 6, 1992, pp. 129–152.
26. A.P. Dempster, N.M. Laird, and D.B. Rubin, "Maximum Likelihood from Incomplete Data via the EM Algorithm," *J. Royal Statist. Society (B)*, vol. 39, 1977, pp. 1–38.
27. J.-T. Chien and J.-C. Junqua, "Unsupervised Hierarchical Adaptation Using Reliable Selection of Cluster-Dependent Parameters," *Speech Communication*, vol. 30, no. 4, 2000, pp. 235–253.

Jen-Tzung Chien was born in Taipei, Taiwan, R.O.C., on August 31, 1967. He received the Ph.D. degree in electrical engineering from the National Tsing Hua University (NTHU), Hsinchu, Taiwan, in 1997.

He was an Instructor in the Department of Electrical Engineering, NTHU, in 1995. In 1997, he joined the Department of Computer Science and Information Engineering, National Cheng Kung University (NCKU), Tainan, Taiwan, where he was an Assistant Professor. Since 2001, he has been an Associate Professor in NCKU. He was the Visiting Researchers at the Speech Technology Laboratory, Panasonic Technologies Inc., Santa Barbara, CA, in 1998, the Department of Computer Science, Tokyo Institute of Technology, Tokyo, Japan, in 2002, and the Institute of Information Science, Academia Sinica, Taiwan, in 2003. His research interests include statistical pattern recognition, speech recognition, face detection/recognition/verification, text mining, biometrics, document classification, multimedia information retrieval, and multimodel human-computer interaction. He serves on the board of the International Journal of Speech Technology and a guest editor of the Special Issue on Chinese Spoken Language Technology.

Dr. Chien is member of the IEEE Signal Processing Society, the International Speech Communication Association, the Chinese Image Processing and Pattern Recognition Society and the Association for Computational Linguistics and Chinese Language Processing (ACLCLP). He serves on the board of ACLCLP. He was listed in Who's Who in the World, 2002, and awarded the Outstanding Young Investigator Award (Ta-You Wu Memorial Award) from National Science Council, Taiwan, 2003.
jtchien@mail.ncku.edu.tw

Jain-Ray Lai received the M.S. degree in computer science and information engineering from National Cheng Kung University, Tainan, Taiwan, in 2000.

His M.S. thesis involves research on robust speech recognition using a microphone array.

Journal of VLSI Signal Processing 36, 153–159, 2004
© 2004 Kluwer Academic Publishers.

Multimedia Corpus of In-Car Speech Communication

NOBUO KAWAGUCHI, KAZUYA TAKEDA AND FUMITADA ITAKURA
*Center for Integrated Acoustic Information Research, Nagoya University, Furo-cho, Chikusa-ku, Nagoya
464-8603 Japan*

Received November 6, 2001; Revised July 11, 2002; Accepted July 22, 2002

Abstract. An ongoing project for constructing a multimedia corpus of dialogues under the driving condition is reported. More than 500 subjects have been enrolled in this corpus development and more than 2 gigabytes of signals have been collected during approximately 60 minutes of driving per subject. Twelve microphones and three video cameras are installed in a car to obtain audio and video data. In addition, five signals regarding car control and the location of the car provided by the Global Positioning System (GPS) are recorded. All signals are simultaneously recorded directly onto the hard disk of the PCs onboard the specially designed data collection vehicle (DCV). The in-car dialogues are initiated by a human operator, an automatic speech recognition (ASR) system and a wizard of OZ (WOZ) system so as to collect as many speech disfluencies as possible.

In addition to the details of data collection, in this paper, preliminary results on intermedia signal conversion are described as an example of the corpus-based in-car speech signal processing research.

1. Introduction

Providing a human-machine interface in a car is one of the most important applications of speech signal processing, where the conventional input/output methods are unsafe and inconvenient. To develop an advanced in-car speech interface, however, not only one but many of the real-world problems, such as noise robustness, distortion due to distant talking [1] and disfluency while driving, must be overcome.

In particular, the difficulty of in-car speech processing is characterized by its variety. The road and traffic conditions, the car condition and the driving movement of the driver change continuously and affect the driver's speech [2]. Therefore, a large corpus is indispensable in the study of in-car speech, not only for training acoustic models under various background noise conditions, [3, 4] but also to build a new model of the combined distortions of speech.

In order to keep pace with the ever-changing environment, it may be helpful to make use of various observed signals rather than to use the speech input signal alone. Therefore, to develop advanced speech process-

ing for in-car application, we need a corpus (1) that covers a large variety of driving conditions, and (2) from which we can extract the conditions surrounding the driver. Constructing such an advanced in-car speech corpus is the goal of the project described in this paper.

In ongoing data collection, a specially built data collection vehicle (DCV) has been used for synchronous recording of seven-channel audio signals, three-channel video signals and vehicle-related signals. About 1 terabytes of data has been collected by recording three sessions of spoken dialogue in about 60 minute of driving for each of 500 drivers. Speech data for text read aloud has also been collected.

In the next section, we describe the DCV which was specially designed for the multichannel audio-visual data acquisition and storage. In Section 3, we present the details of the data collected and the methodology used for data collection. In Section 4, the spoken dialogue system that utilizes automatic speech recognition capability for evaluating the actual performance of the speech recognizer in an in-car environment is introduced. Finally, preliminary studies

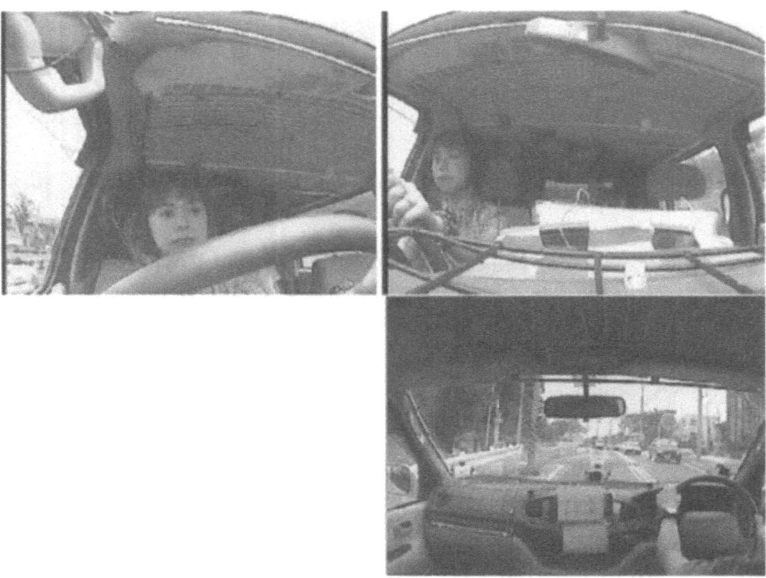

Figure 1. Visual signal captured by the three cameras. (a) The driver's face (left upper), (b) the driver, the operator and the back view (right upper) and (c) front view (right bottom).

on intermedia data conversion will be described as an example of the corpus study, in Section 5.

2. Data Collection Vehicle

The DCV is a car specially designed for the collection of multimedia data. The vehicle is equipped with eight network-connected personal computers (PCs). Three PCs have a 16-channel analog-to-digital and digital-to-analog conversion port that can be used for recording and playing back data. The data can be digitized using 16-bit resolution and sampling frequencies up to 48 kHz. One of these three PCs can be used for recording audio signals from 16 microphones. The second PC can be used for audio playback on 16 loudspeakers. The third PC is used for recording five signals associated with the vehicle: the angle of the steering wheel, the status of the accelerator and brake pedals, the speed of the car and the engine speed. These vehicle-related data are recorded at a sampling frequency of 1 kHz in 2-byte resolution. In addition, location information obtained from the Global Positioning System (GPS) is also recorded at the sampling frequency of 1 Hz.

Three other PCs are used for recording video images (Fig. 1). The first camera captures the face of the driver. The second camera captures the conversation between the driver and the experiment navigator. The third camera captures the view through the windshield. These images are coded into MPEG1 format. The remaining two PCs are used for controlling the experiment. The multimedia data on all systems are recorded synchronously. The total amount of data is about 2 gigabytes for about a 60-minute drive during which three dialogue sessions are recorded. The recorded data is directly stored on the hard disks of the PCs in the car.

Figure 2 shows the arrangement of equipment in the DCV, including the PCs, a power generator with batteries, video controller, microphone amplifiers and speaker amplifiers. An alternator and a battery are installed for stabilizing the power supply. Figure 3 shows

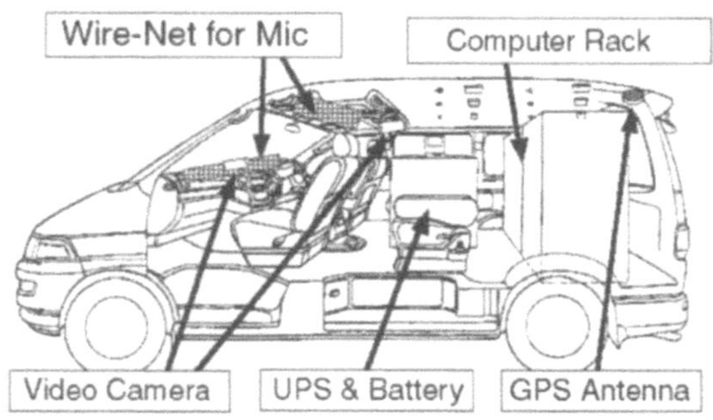

Figure 2. Configuration of DCV.

Table 1. Specifications of recording devices.

Type of data	Specifications
Sound input	16 ch, 16 bit, 16 kHz
Sound output	16 ch, 16 bit, 16 kHz
Video input	3 ch, MPEG1
Control signal	Status of accelerator and brake, angle of steering wheel Engine RPM, Speed: 16 bit, 1 kHz
Location	Differential Global Positioning System (DGPS)

Figure 3. The interior of the DCV. Wire nets are attached for various arrangement of microphones.

the interior of the DCV. Wire nets are attached to the ceiling of the car so that the microphones can be arranged in arbitrary positions. Figure 4 shows plots of the vehicle-related data such as the status of brake and accelerator pedals, the rotational speed of the engine motor, and the speed.

3.   Speech Materials

The collected speech materials are listed in Table 2. The task domain of the dialogues is the restaurant guidance around the Nagoya University campus. In dialogues with a human operator and the WOZ system, we have prompted the driver to issue natural and varied utterances related to the task domain, by displaying a 'prompt panel'. On the panel, a keyword, such as *fast food*, *bank*, *Japanese food*, or *parking*, or a situation sentence, such as 'Today is an anniversary. Let's have a party.', 'I am so hungry. I need to eat!' or 'I am thirsty. I want a drink!', are displayed. In these modes, therefore, the driver takes the initiative in the dialogue. The

Table 2. Speech materials recorded in the experiment.

Item	Approx. time
Prompted dialogue	5 min
Natural dialogue	5 min
Dialogue with system	5 min
Dialogue with WOZ	5 min
Repeating phonetically balanced sentences (driving)	10 min
Reading phonetically balanced sentences (idling)	5 min

operator also navigates the driver to a predetermined destination while they are having a dialogue, in order to simulate the common function of a car-navigation system. All responses of the operator are given by synthetic speech in the WOZ mode. In addition, fully natural dialogues are also conducted between the driver and a distant operator via cellular phone. In such natural dialogues, the driver asks for the telephone number of a shop from the yellow pages information service. These natural dialogues are collected both when idling the engine and while driving the car.

All utterances have been phonetically transcribed and tagged with time codes. Tagging is performed separately for utterances by the driver and by the operator so that timing analysis of the utterances can be carried out. On average, there are 380 utterances and 2768 morphemes in the data for a driver.

In addition to the dialogues, speech of the text read aloud and isolated word utterances have also been collected. Each subject read 100 phonetically balanced sentences while idling the engine and 25 sentences while driving the car. A speech prompter is used to present the text while driving. The speech data of the read text is mainly used for training acoustic models. The set of isolated word utterances consists of digit strings and car control words.

4.   Preliminary Results on Collected Data

Since dialogue between man and machine is one of our final goals, we are collecting man-machine dialogues using a prototype spoken dialogue system that has speech recognition capabilities. The task domain of the prototype system is restaurant information. Drivers can retrieve information and make a reservation at a restaurant near the campus by conversing with the system. The automatic speech recognition module of the system is based on a common dictation software platform known as Julius 3.1 [8].

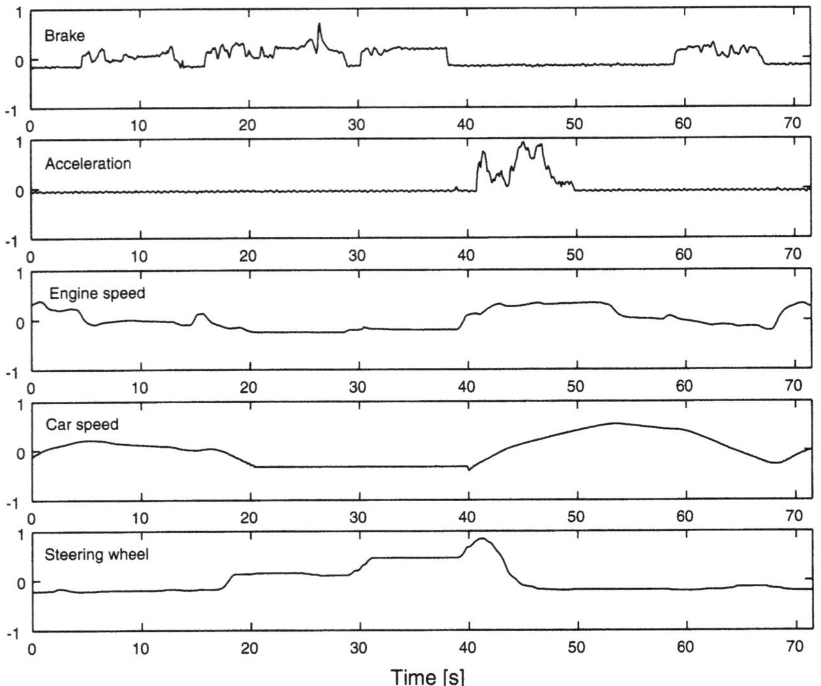

Figure 4. Vehicle-related signals. Brake and accelerator pedals, rotational speed of the engine motor and speed (from top to bottom).

A trigram language model with a 1500-word vocabulary is trained using about 10000 sentences. The main body of the training sentences is extracted from the human-human dialogue collected in the early stage of the experiment. The other sentences are generated from a finite state grammar that accepts permissible utterances in the task domain. State clustered triphone hidden Markov models consisting of 3000 states are used as acoustic models. The number of mixtures for each state is 16. The models are trained using 40,000 phonetically balanced sentences uttered by 200 speakers recorded in a soundproof room with a close-talking microphone [9]. The same microphone as in this recording is used for speech input in the prototype dialogue system. A preliminary evaluation of the speech recognition module of the system over 1,153 sentences of 48 speakers has given a word correctness of about 70% under real driving conditions in the city area, where the typical driving speed is less than 40 km/h.

The dialogue is controlled by transitions among 12 states, each of which corresponds to the database query results. When a set of particular conditions defined for a transition is satisfied, the predefined state transition occurs, invoking associated actions, i.e., generating speech responses. Examples of the dialogue states are listed in the Table 3. Up to today, 75% of the man-machine dialogues have been correctly completed by the system.

An example of a dialogue between the ASR system and a user is shown in Fig. 5.

5. Cross-Media Signal Conversion

Among the ongoing research projects concurrent to data collection, cross-media signal conversion makes use of the multimedia aspect of the corpus. In this

Table 3. Examples of the states defined for controlling dialogues.

ID	Function
1	Initial state.
2	More than three restaurants to announce
3	No restaurant to announce
4	Less than three restaurants to announce
5	The restaurant is decided
6	Error (the dialogue has too many turns.)
7	Error (the session is too long.)
8	Task completed

```
System: Please give me a condition for restaurant query.
User: Japanese restaurant that serves about 1000-yen food.
System: Japanese, 1000-yen. Is that right ?
User: Yes.
System: There are Mikado, Shogun and Kyoto near here.
System: Which restaurant do you prefer ?
User: I would like to go to Mikado.
System: Mikado. Is that right ?
User: Yes.
System: You can reserve a table for that restaurant.
System: Do you reserve ?
User: No.
System: Do you need a navigation to the restaurant ?
User: No.
```

Figure 5.   An example of a dialogue between the ASR system and a subject.

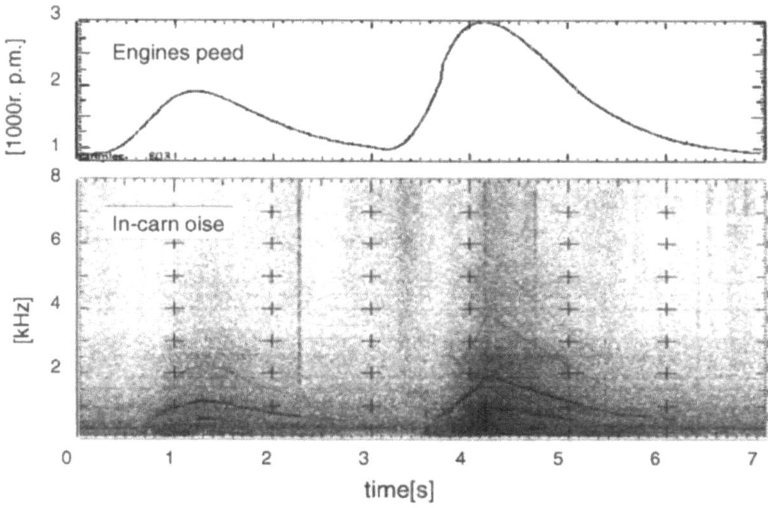

Figure 6.   Engine speed and related in-car noise.

section, a typical example of the research effort, synthesizing in-car noise, is described.

Because the nonstationarity of the background noise is one of the most difficult problems in in-car speech processing, accurate prediction of the noise signal is expected to improve various speech processing performances. In this study, therefore, in-car noise is predicted using the driving signals, i.e., the car speed and the engine speed signals. As shown in Fig. 6, the engine speed governs the timbre of in-car noise by producing peaks in the spectrum. The power contour of the noise signal, on the other hand, is proportional to the car speed signal. Therefore, one can model the car noise signal as a superposition of the engine-

speed-dependent and car-speed-dependent noise signals. Hereafter, these two noise signals are referred to as engine noise and road noise. Because both the engine-speed and car-speed signals are collected synchronously with the in-car noise signal, we can formulate the noise prediction problem as cross-media signal conversions, i.e., conversion from the engine-speed and car-speed signals to the in-car noise signal.

The engine-speed-dependent signal is synthesized by convoluting random white Gaussian noise with the frequency response of the given engine speed. The frequency response is calculated by adding some spectral peaks at the fundamental frequency, f_c, given as 16 times the combustion frequency, and their multiples

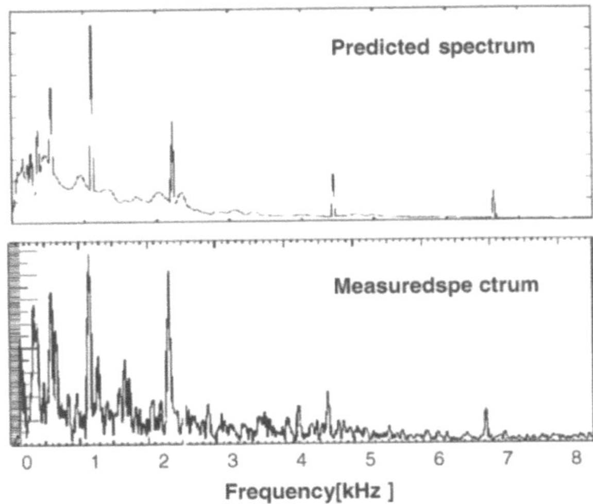

Figure 7. Spectrum of the synthesized (above) and the original (below) engine noise.

to the averaged spectral magnitude $\langle H(f) \rangle$, i.e.,

$$H(f) = \langle H(f) \rangle + A\delta(f - f_c)$$
$$+ \sum_{n=1}^{3} \frac{A}{n+1}\delta(f - 2nf_c)$$
$$+ \sum_{n=2}^{4} \frac{A}{n}\delta(f - 1/nf_c),$$

where $\delta(f - f_c)$ represents a line spectrum at frequency f_c ([Hz]). Therefore, the third term of the right hand side of the equation gives the harmonics of the fundamental frequency, whereas the fourth term gives the subharmonic components, which arise due to the nonlinearity of the vibration. The validity of the model can be seen in Fig. 7 where the predicted spectrum is seen to simulate the measured spectrum well.

The road noise, on the other hand, is generated by concatenating speed-dependent noise waveform prototypes. A noise waveform prototype is prepared for every 5 km/h speed, i.e., 5, 10, ..., 85 km/h. For the generation of the noise signal, the prototypes are replaced every 2 seconds according to the car speed.

In a subjective test, the mean opinion score (MOS) of the generated noise signal was rated 3.1, whereas the recorded noise and white noise, whose amplitude envelope is proportional to the car speed, were rated 4.6 and 1.1, respectively. These results confirm the feasibility of cross-media signal conversion.

6.   Summary

In this paper, we presented details of constructing a multimedia corpus of in-car speech communication. The corpus consists of synchronously recorded multi-channel audio/video signals, driving signals and GPS output. The spoken dialogues of the driver were collected in various styles, i.e., human-human and human-machine, prompted and natural, for the restaurant guidance task domain. An ASR system was utilized for collecting human-machine dialogues.

To date, more than 500 subjects have been enrolled in data collection. Ongoing acoustic/language model training for robust speech recognition is being performed using the corpus. Furthermore, making full use of the multimedia characteristics of the corpus, a cross-media signal conversion study has begun.

The corpus will be available for various research purposes in early 2002, and is planned to be extended to English and other languages.

Acknowledgments

This research has been supported by a Grant-in-Aid for COE Research (No. 11CE2005).

References

1. J.C. Junqua and J.P. Haton, *Robustness in Automatic Speech Recognition*. Kluwer Academic Publishers, 1996.
2. D. Roy, "'Grounded' Speech Communication," in *Proc. of the International Conference on Spoken Language Processing, ICSLP 2000*, Beijin, 2000, pp. IV69–IV72
3. P. Gelin and J.C. Junqua, "Techniques for Robust Speech Recognition in the Car Environment," in *Proc. of European Conference Speech Communication and Technology, EUROSPEECH'99*, Budapest, 1999.
4. M.J. Hunt, "Some Experiences in In-Car Speech Recognition," in *Proc. of the Workshop on Robust Methods for Speech Recognition in Adverse Conditions*, Tampere, 1999, pp. 25–31
5. P. Geutner, L. Arevalo, and J. Breuninger, "VODIS—Voice-Operated Driver Information Systems: A Usability Study on Advanced Speech Technologies for Car Environments," in *Proc. of International Conference on Spoken Language Processing*, ICSLP2000, Beijin, 2000, pp. IV378–IV381.
6. A. Moreno, B. Lindberg, C. Draxler, G. Richard, K. Choukri, J. Allen, and Stephan Eule, "SpeechDat-Car: A Large Speech Database for Automotive Environments," in *Proc. of 2nd Int'l Conference on Language Resources and Evaluation*, Athens, LREC 2000.
7. N. Kawaguchi, S. Matsubara, H. Iwa, S. Kajita, K. Takeda, F. Itakura, and Y. Inagaki, "Construction of Speech Corpus in Moving Car Environment," in *Proc. of International Conference on Spoken Language Processing*, ICSLP2000, Beijin, 2000 pp. 362–365.

8. T. Kawahara, T. Kobayashi, K. Takeda, N. Minematsu, K. Itou, M. Yamamoto, A. Yamada, T. Utsuro, and K. Shikano, "Japanese Dictation Toolkit: Plug-and-Play Framework for Speech Recognition R&D," in *Proc. of IEEE Automatic Speech Recognition and Understanding Workshop (ASRU'99)*, 1999 pp. 393–396.

9. K. Itou, M. Yamamoto, K. Takeda, T. Takezawa, T. Matsuoka, T. Kobayashi, K. Shikano, and S. Itahashi, JNAS: Japanese Speech Corpus for Large Vocabulary Continuous Speech Recognition Research, *J. Acoust. Soc. Jpn.(E)*, vol. 20, no. 3, 1999, pp. 199–206.

Journal of VLSI Signal Processing 36, 161–187, 2004
© 2004 Kluwer Academic Publishers.

Speech and Language Processing for Multimodal Human-Computer Interaction*

L. DENG, Y. WANG, K. WANG, A. ACERO, H. HON, J. DROPPO, C. BOULIS, M. MAHAJAN
AND X.D. HUANG
Microsoft Research, One Microsoft Way, Redmond, WA 98052, USA

Received November 20, 2001; Revised June 18, 2002; Accepted June 18, 2002

Abstract.   In this paper, we describe our recent work at Microsoft Research, in the project codenamed *Dr. Who*, aimed at the development of enabling technologies for speech-centric multimodal human-computer interaction. In particular, we present in detail MiPad as the first *Dr. Who's* application that addresses specifically the mobile user interaction scenario. MiPad is a wireless mobile PDA prototype that enables users to accomplish many common tasks using a multimodal spoken language interface and wireless-data technologies. It fully integrates continuous speech recognition and spoken language understanding, and provides a novel solution to the current prevailing problem of pecking with tiny styluses or typing on minuscule keyboards in today's PDAs or smart phones. Despite its current incomplete implementation, we have observed that speech and pen have the potential to significantly improve user experience in our user study reported in this paper. We describe in this system-oriented paper the main components of MiPad, with a focus on the robust speech processing and spoken language understanding aspects. The detailed MiPad components discussed include: distributed speech recognition considerations for the speech processing algorithm design; a stereo-based speech feature enhancement algorithm used for noise-robust front-end speech processing; Aurora2 evaluation results for this front-end processing; speech feature compression (source coding) and error protection (channel coding) for distributed speech recognition in MiPad; HMM-based acoustic modeling for continuous speech recognition decoding; a unified language model integrating context-free grammar and N-gram model for the speech decoding; schema-based knowledge representation for the MiPad's personal information management task; a unified statistical framework that integrates speech recognition, spoken language understanding and dialogue management; the robust natural language parser used in MiPad to process the speech recognizer's output; a machine-aided grammar learning and development used for spoken language understanding for the MiPad task; *Tap & Talk* multimodal interaction and user interface design; back channel communication and MiPad's error repair strategy; and finally, user study results that demonstrate the superior throughput achieved by the *Tap & Talk* multimodal interaction over the existing pen-only PDA interface. These user study results highlight the crucial role played by speech in enhancing the overall user experience in MiPad-like human-computer interaction devices.

Keywords: speech-centric multimodal interface, human-computer interaction, robust speech recognition, SPLICE algorithm, denoising, online noise estimation, distributed speech processing, speech feature encoding, error protection, spoken language understanding, automatic grammar learning, semantic schema, user study

*Various portions of this article have been presented at ICSLP-2000 [1–3] in Beijing, China, ICASSP-2001 [4–6] in Salt Lake City, US, Eurospeech-1999 [7] in Hungary, Eurospeech-2001 [8] in Denmark, and at ASRU-2001 Workshop [9, 10] in Italy.

1. Introduction

When humans converse and interact with one another, we often utilize a wide range of communicative devices and effectively exploit these devices to clearly and concisely express our communicative intentions. In contrast, in human-computer interaction, humans have typically been limited to interactions through the use of the graphical user interface (GUI) including a graphical display, a mouse, and a keyboard. While GUI-based technology has significantly improved human computer interface by using intuitive real-world metaphors, it nevertheless still makes human's interaction with computers far from natural and flexible. Intensive research under the general name of multimodal interfaces has been actively pursued in recent years. These interfaces provide the potential of allowing users to interact with computers through several different modalities such as speech, graphical displays, keypads, and pointing devices, and offer significant advantages over traditional unimodal interfaces. The purpose of this paper is to present some of our group's recent research activities, under the project's codename *Dr. Who*, on speech and language processing as applied to speech-centric multimodal human-computer interaction. In particular, we will describe the first showcase project for *Dr. Who*, the Multimodal Intelligent Personal Assistant Device, or MiPad, which is a next generation Personal Digital Assistant (PDA) prototype embedding several key sub-areas of spoken language technology. These sub-areas include speech analysis, environment-robust speech processing, spoken language understanding, and multimodal interaction—all directly pertinent to the major themes in this special issue on real-world speech processing.

One particular limitation of the traditional GUI which we specifically address in our research is the heavy reliance of GUI on a sizeable screen, keyboard and pointing device, whereas such a screen or device is not always available. With more and more computers being designed for mobile usages and hence being subject to the physical size and hands-busy or eyes-busy constraints, the traditional GUI faces an even greater challenge. Multimodal interface enabled by spoken language is widely believed to be capable of dramatically enhancing the usability of computers because GUI and speech have complementary strengths. While spoken language has the potential to provide a natural interaction model, the ambiguity of spoken language and the memory burden of using speech as output modality on the user have so far prevented it from becoming the choice of a mainstream interface. MiPad is one of our attempts in developing enabling technologies for speech-centric multimodal interface. MiPad is a prototype of wireless PDA that enables users to accomplish many common tasks using a multimodal spoken language interface (speech + pen + display). One of MiPad's hardware design concepts is illustrated in Fig. 1.

MiPad intends to alleviate the current prevailing problem of pecking with tiny styluses or typing on minuscule keyboards in today's PDAs by introducing the speech capability using a built-in microphone. However, resembling more a PDA and less a telephone, MiPad intentionally avoids speech-only interactions. MiPad is designed to support a variety of tasks such as

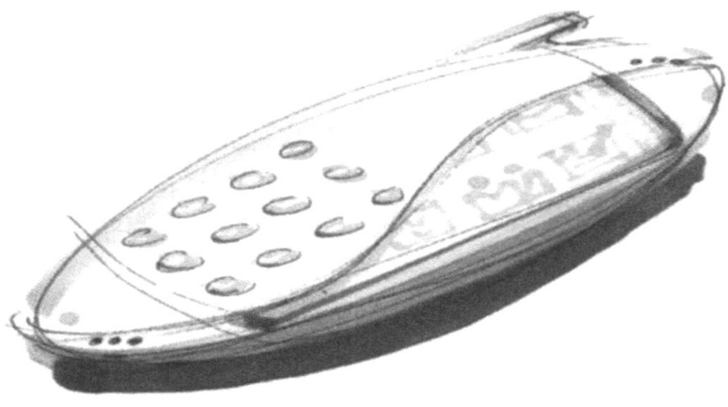

Figure 1. One of MiPad's industrial design template.

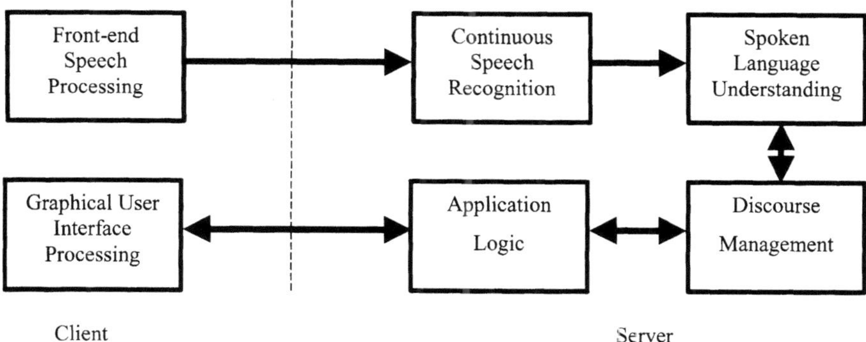

Figure 2. MiPad's client-server (peer-to-peer) architecture. The client is based on a W ndows CE iPAQ, and the server is based on a Windows server. The client-server communication is currently based on the wireless LAN.

E-mail, voice-mail, calendar, contact list, notes, web browsing, mobile phone, and document reading and annotation. This collection of functions unifies the various devices that people carry around today into a single, comprehensive communication and productivity tool. While the entire functionality of MiPad can be accessed by pen alone, it is preferred to be accessed by speech and pen combined. The user can dictate to an input field by holding the pen down in it. Alternatively, the user can also select the speech field by using the roller to navigate and by holding it down while speaking. The field selection, called *Tap & Talk*, not only indicates where the recognized text should go but also serves as a push to talk control. *Tap & Talk* narrows down the number of possible instructions for spoken language processing. For example, selecting the *"To:"* field on an e-mail application display indicates that the user is about to enter a name. This dramatically reduces the complexity of spoken language processing and cuts down the speech recognition and understanding errors to the extent that MiPad can be made practically usable despite the current well known limitations of speech recognition and natural language processing technology.

One unique feature of MiPad is a general purpose "Command" field to which a user can issue naturally spoken commands such as "Schedule a meeting with Bill tomorrow at two o'clock." From the user's perspective, MiPad recognizes the commands and follows with necessary actions. In response, it pops up a "meeting arrangement" page with related fields (such as date, time, attendees, etc.) filled appropriately based on the user's utterance. Up to this date, MiPad fully implements PIM (Personal Information Management) functions including email, calendar, notes, task, and contact list with a hardware prototype based on Compaq's iPaq PDA. All MiPad applica-

tions are configured in a client-server architecture as shown in Fig. 2. The client on the left side of Fig. 2 is MiPad powered by Microsoft Windows CE operating system that supports (1) sound capturing ability, (2) front-end acoustic processing including noise reduction, channel normalization, feature compression, and error protection, (3) GUI processing module, and (4) a fault-tolerant communication layer that allows the system to recover gracefully from the network connection failures. Specifically, to reduce bandwidth requirements, the client compresses the wideband speech parameters down to a maximal 4.8 Kbps bandwidth. Between 1.6 and 4.8 Kbps, we observed virtually no increase in the recognition error on some tasks tested. A wireless local area network (WLAN), which is currently used to simulate a third generation (3G) wireless network, connects MiPad to a host machine (server) where the continuous speech recognition (CSR) and spoken language understanding (SLU) take place. The client takes approximately 450 KB of program space and an additional 200 KB of runtime heap, and utilizes approximately 35% of CPU time with iPAQ's 206 MHz StrongARM processor. At the server side, as shown on the right side of Fig. 2, MiPad applications communicate with the continuous speech recognition (CSR) and spoken language understanding (SLU) engines for coordinated context-sensitive *Tap & Talk* interaction.

Given the above brief overview of MiPad which exemplifies our speech-centric multimodal human-computer interaction research, the remaining of this paper will describe details of MiPad with an emphasis on speech processing, spoken language understanding, and the UI design considerations. Various portions of this paper have been presented at the several conferences (ICSLP-2000 [1–3], ICASSP-2001 [4–6],

Eurospeech-1999 [7], Eurospeech-2001 [8] and ASRU-2001 Workshop [9, 10]). The purpose of this paper is to synthesize these earlier presentations on the largely isolated MiPad components into a single coherent one so as to highlight the important roles of speech and language processing in the MiPad design, and report some most recent research results. The organization of this paper is as follows. In Sections 2 and 3, we describe our recent work on front-end speech processing, including noise robustness and source/channel coding, which underlie MiPad's distributed speech recognition capabilities. Acoustic models and language models used for the decoding phase of MiPad continuous speech recognition are presented in Section 4. Spoken language understanding using the domains related to MiPad and other applications is described in Section 4, together with dialogue management directly linked to the application logic and user interface processing. Finally, MiPad user interface and user study results are outlined in Section 4, and a summary is provided in Section 7.

2. Robust Speech Processing

Robustness to acoustic environment or immunity to noise and channel distortion is one most important aspect of MiPad design considerations. For MiPad to be acceptable to the general public, it is desirable to remove the need for a close-talking microphone in capturing speech. Although close-talking microphones pick up relatively little background noise and allow speech recognizers to achieve high accuracy for the MiPad-domain tasks, it is found that users much prefer built-in microphones even if there is minor accuracy degradation. With the convenience of using built-in microphones, noise robustness becomes a key challenge to maintaining desirable speech recognition and understanding performance. Our recent work on acoustic modeling aspects of MiPad has focused mainly on this noise-robustness challenge. In this section we will present most recent results in the framework of distributed speech recognition (DSR) that MiPad design has adopted.

2.1. Distributed Speech Recognition Considerations for Algorithm Design

There has been a great deal of interest recently in standardizing DSR applications for a plain phone, PDA,

or a smart phone where speech recognition is carried out at a remote server. To overcome bandwidth and infrastructure cost limitations, one possibility is to use a standard codec on the device to transmit the speech to the server where it is subsequently decompressed and recognized. However, since speech recognizers such as the one in MiPad only need some features of the speech signal (e.g., Mel-cepstrum), bandwidth can be further saved by transmitting only those features. Much of the recent research programs in the area of DSR have concentrated on the Aurora task [11], representing a concerted effort to standardize a DSR front-end and to address the issues surrounding robustness to noise and channel distortions at a low bit rate. Our recent work on noise robustness in DSR has been concentrated on the Aurora task [8].

In DSR applications, it is easier to update software on the server because one cannot assume that the client is always running the latest version of the algorithm. With this consideration in mind, while designing noise-robust algorithms for MiPad, we strive to make the algorithms front-end agnostic. That is, the algorithms should make no assumptions on the structure and processing of the front end and merely try to undo whatever acoustic corruption that has been shown during training. This consideration also favors approaches in the feature rather than the model domain.

Here we describe one particular algorithm that has so far given the best performance on the Aurora2 task and other Microsoft internal tasks with much larger vocabularies. We called the algorithm SPLICE, short for Stereo-based Piecewise Linear Compensation for Environments. In a DSR system, the SPLICE may be applied either within the front end on the client device, or on the server, or on both with collaboration. Certainly a server side implementation has some advantages as computational complexity and memory requirements become less of an issue and continuing improvements can be made to benefit even devices already deployed in the field. Another useful property of SPLICE in the serve implementation is that new noise conditions can be added as they are identified once by a server. This can make SPLICE quickly adapt to any new acoustic environment with minimum additional resource.

2.2. Basic Version of the SPLICE Algorithm

SPLICE is a frame-based, bias removal algorithm for cepstrum enhancement under additive noise, channel distortion or a combination of the two. In [1] we reported the approximate MAP (Maximum A Posteriori) formulation of the algorithm, and more recently in [4, 8] we described the MMSE (Minimum Mean Square Error) formulation of the algorithm with a much wider range of naturally recorded noise, including both artificially mixed speech and noise, and naturally recorded noisy speech.

SPLICE assumes no explicit noise model, and the noise characteristics are embedded in the piecewise linear mapping between the "stereo" clean and distorted speech cepstral vectors. The piecewise linearity is intended to approximate the true nonlinear relationship between the two. The nonlinearity between the clean and distorted (including additive noise) cepstral vectors arises due to the use of the logarithm in computing the cepstra. Because of the use of the stereo training data that provides accurate estimates of the bias or correction vectors without the need for an explicit noise model, SPLICE is potentially able to handle a wide range of distortions, including nonstationary distortion, joint additive and convolutional distortion, and nonlinear distortion (in time-domain). One key requirement for the success of the basic version of SPLICE described here is that the distortion conditions under which the correction vectors are learned from the stereo data must be similar to those corrupting the test data. Enhanced versions of the algorithm described later in this section will relax this requirement by employing a noise estimation and normalization procedure. Also, while the collection of a large amount of stereo data may be inconvenient for implementation, the collection is needed only in the training phase. Further, our experience suggests that a relatively small amount of stereo data is already sufficient for the SPLICE training, especially for the noise normalization version of the SPLICE.

We assume a general nonlinear distortion of a clean cepstral vector, $\mathbf{x}$, into a noisy one, $\mathbf{y}$. This distortion is approximated in SPLICE by a set of linear distortions. The probabilistic formulation of the basic version of SPLICE is provided below.

2.2.1. Basic Assumptions.
The first assumption is that the noisy speech cepstral vector follows a mixture distribution of Gaussians

$$p(\mathbf{y}) = \sum_s p(\mathbf{y} \mid s)p(s), \quad \text{with}$$

$$p(\mathbf{y} \mid s) = N(\mathbf{y}; \boldsymbol{\mu}_s, \boldsymbol{\Sigma}_s) \tag{1}$$

where s denotes the discrete random variable taking the values $1, 2, \ldots, N$, one for each region over which the piecewise linear approximation between the clean cepstral vector $\mathbf{x}$ and distorted cepstral vector is made. This distribution, one for each separate distortion condition (not indexed for clarity), can be thought as a "codebook" with a total of N codewords (Gaussian means) and their variances.

The second assumption made by SPLICE is that the conditional probability density function (PDF) for the clean vector $\mathbf{x}$ given the noisy speech vector, $\mathbf{y}$, and the region index, s, is a Gaussian with the mean vector to be a linear combination of the noisy speech vector $\mathbf{y}$. In this paper, we take a simplified form of this (piecewise) function by making the rotation matrix to be identity one, leaving only the bias or correction vector. Thus, the conditional PDF has the form

$$p(\mathbf{x} \mid \mathbf{y}, s) = N(\mathbf{x}; \mathbf{y} + \mathbf{r}_s, \boldsymbol{\Gamma}_s) \tag{2}$$

2.2.2. SPLICE Training.
Since the noisy speech PDF $p(\mathbf{y})$ obeys mixture-of-Gaussian distribution, the standard EM (Expectation and Maximization) algorithm is used to train $\boldsymbol{\mu}_s$ and $\boldsymbol{\Sigma}_s$. Initial values of the parameters can be determined by a VQ (Vector Quantization) clustering algorithm.

The parameters $\mathbf{r}_s$ and $\boldsymbol{\Gamma}_s$ of the conditional PDF $p(\mathbf{x} \mid \mathbf{y}, s)$ can be trained using the maximum likelihood criterion. Since the variance of the distribution is not used in cepstral enhancement, we only give the ML estimate of the correction vector below:

$$\mathbf{r}_s = \frac{\sum_n p(s \mid \mathbf{y}_n)(\mathbf{x}_n - \mathbf{y}_n)}{\sum_n p(s \mid \mathbf{y}_n)}, \tag{3}$$

where

$$p(s \mid \mathbf{y}_n) = \frac{p(\mathbf{y}_n \mid s)p(s)}{\sum_s p(\mathbf{y}_n \mid s)}. \tag{4}$$

This training procedure requires a set of stereo (two channel) data. One channel contains the clean utterance, and the other channel contains the same utterance with distortion, where the distortion represented by the correction vectors is estimated above. The two-channel data can be collected, for example, by simultaneously recording utterances with one close-talk and one far-field microphone. Alternatively, it has been shown in our research that a large amount of synthetic data can be bootstrapped from a small amount of real data with virtually no loss of speech recognition accuracy.

2.2.3. SPLICE for Cepstral Enhancement. One significant advantage of the above two basic assumptions made in SPLICE is the inherent simplicity in deriving and implementing the rigorous MMSE estimate of clean speech cepstral vectors from their distorted counterparts. Unlike the FCDCN (Fixed Code-Dependent Cepstral Normalization) algorithm [1], no approximations are made in deriving the optimal enhancement rule. The derivation is outlined below.

The MMSE is the following conditional expectation of clean speech vector given the observed noisy speech:

$$E_x[\mathbf{x} \mid \mathbf{y}] = \sum_s p(s \mid \mathbf{y}) E_\mathbf{x}[\mathbf{x} \mid \mathbf{y}, s]. \tag{5}$$

Due to the second assumption of SPLICE, the above codeword-dependent conditional expectation of $\mathbf{x}$ (given $\mathbf{y}$ and s) is simply the bias-added noisy speech vector:

$$E_x[\mathbf{x} \mid \mathbf{y}, s] = \mathbf{y} + \mathbf{r}_s \tag{6}$$

where bias $\mathbf{r}_s$ has been estimated from the stereo training data according to Eq. (3). This gives the simple form of the MMSE estimate as the noisy speech vector corrected by a linear weighted sum of all codeword-dependent bias vectors already trained:

$$\hat{\mathbf{x}} = E_x[\mathbf{x} \mid \mathbf{y}] = \mathbf{y} + \sum_s p(s \mid \mathbf{y}) \mathbf{r}_s \tag{7}$$

While this is already efficient to compute, more efficiency can be achieved by approximating the weights according to

$$\hat{p}(s \mid \mathbf{y}) \cong \begin{cases} 1 & s = \arg\max_s p(s \mid \mathbf{y}) \\ 0 & \text{otherwise} \end{cases}. \tag{8}$$

This approximation turns the MMSE estimate to the approximate MAP estimate that consists of two sequential steps of operation. First, finding optimal codeword s using the VQ codebook based on the parameters $(\mathbf{r}_s, \Gamma_s)$, and then adding the codeword-dependent vector $\mathbf{r}_s$ to the noisy speech vector. We have found empirically that the above VQ approximation does not appreciably affect recognition accuracy.

2.3.   Enhancing SPLICE by Temporal Smoothing

In this enhanced version of SPLICE, we not only minimize the static deviation from the clean to noisy cepstral vectors (as in the basic version of SPLICE), but also seeks to minimize the dynamic deviation.

The basic SPLICE optimally processes each frame of noisy speech independently. An obvious extension is to jointly process a segment of frames. In this way, although the deviation from the clean to noisy speech cepstra for an individual frame may be undesirably greater than that achieved by the basic, static SPLICE, the global deviation that takes into account the differential frames and the whole segment of frames will be reduced compared with the basic SPLICE.

We have implemented the above idea of "dynamic SPLICE" through temporally smoothing the bias vectors obtained from the basic, static SPLICE. This is an empirical way of implementing the rigorous solution via the use of a more realistic model for the time-evolution of the clean speech dynamics, either in the discrete state, $p(\mathbf{x}_n \mid \mathbf{y}_n, s_n, s_{n-1})$, or in the continuous clean speech vector estimate, $p(\mathbf{x}_n \mid \mathbf{y}_n, s_n, \mathbf{x}_{n-1})$.

An efficient way of implementing an approximate dynamic SPLICE is to time-filter each component of the cepstral bias vector $\mathbf{r}_{s_n}$. We have achieved significant performance gains using this efficient heuristic implementation. In our specific implementation, we used a simple zero-phase, non-causal, IIR filter to smooth the cepstral bias vectors. This filter has a low-pass characteristic, with the system transfer function of

$$H(z) = \frac{-0.5}{(z^{-1} - 0.5)(z - 2)}. \tag{9}$$

This transfer function is the result of defining an objective function as the summation of the static and dynamic deviations from clean speech to noisy speech vectors. The optimal solution that minimizes this objective function is of the form of Eq. (9), where the constants are functions of the variances in the speech model. In practice, we found in experiments that using Eq. (9) instead of the exact solution produces similar results at a lower computational cost.

2.4.   Enhancing SPLICE by Noise Estimation and Noise Normalization

One well recognized deficiency of the SPLICE algorithm discussed so far is that the noise condition is often unseen in the collected stereo data used to train the SPLICE codebooks. In the new, noise-normalized version of SPLICE, different noise conditions between the SPLICE training set and test set are normalized.

The procedure for noise normalization and for denoising is as follows. Instead of building codebooks for noisy speech feature vector $\mathbf{y}$ from the training set, they are built from $\mathbf{y} - \mathbf{n}$, where $\mathbf{n}$ is an estimated noise from $\mathbf{y}$. Then the correction vectors are estimated from the training set using the noise-normalized stereo data $(\mathbf{y} - \mathbf{n})$ and $(\mathbf{x} - \mathbf{n})$. The correction vectors trained in this new SPLICE will be different from those in the basic version of SPLICE. This is because the codebook selection will be different since $p(s \mid \mathbf{y})$ is changed to $p(s \mid \mathbf{y} - \mathbf{n})$. For denoising in the test data, the noise-normalized noisy cepstra $\mathbf{y} - \mathbf{n}$ are used to obtain the noise-normalized MMSE estimate, and then the noise normalization is undone by adding the estimated noise $\mathbf{n}$ back to the MMSE estimate. This noise normalization procedure is intended to eliminate possible mismatch between the environment where SPLICE stereo training takes place and the environment where SPLICE is deployed to remove the noise in the test data.

Our research showed that the effectiveness of the above noise-normalized SPLICE is highly dependent on the accuracy of the noise estimate $\mathbf{n}$ [8, 9]. We have carried out research on several ways of automatically estimating nonstationary noise in the Aurora2 database. We describe below one algorithm that has given by far the highest accuracy in noise estimation and at the same time by far the best noise-robust speech recognition results evaluated on the Aurora2 task.

2.5. *Nonstationary Noise Estimation by Iterative Stochastic Approximation*

In [9], a novel algorithm is proposed, implemented, and evaluated for recursive estimation of parameters in a nonlinear model involving incomplete data. The algorithm is applied specifically to time-varying deterministic parameters of additive noise in a mildly nonlinear model that accounts for the generation of the cepstral data of noisy speech from the cepstral data of the noise and clean speech. For computer recognition of the speech that is corrupted by highly nonstationary noise, different observation data segments correspond to very different noise parameter values. It is thus strongly desirable to develop recursive estimation algorithms, since they can be designed to adaptively track the changing noise parameters. One such design based on the novel technique of iterative stochastic approximation within the recursive-EM framework is developed and evaluated. It jointly adapts time-varying

noise parameters and the auxiliary parameters introduced to piecewise linearly approximate the nonlinear model of the acoustic environment. The accuracy of approximation is shown to have improved progressively with more iterations.

The essence the algorithm is the use of iterations to achieve close approximations to a nonlinear model of the acoustic environment while at the same time employing the "forgetting" mechanism to effectively track nonstationary noise. Using a number of empirically verified assumptions associated with the implementation simplification, the efficiency of this algorithm has been improved close to real time for noise tracking. The mathematical theory, algorithm, and implementation detail of this iterative stochastic approximation technique can be found in [9, 12].

Figures 3–5 show the results of noise-normalized SPLICE denoising using the iterative stochastic algorithm for tracking nonstationary noise $\mathbf{n}$ in an utterance of the Aurora2 data, where the SNR is 10 dB, 5 dB, and 0 dB, respectively. From top to bottom panels are noisy speech, clean speech, and denoised speech, all in the same spectrogram format. Most of the noise has been effectively removed, except for some strong noise burst located around 150–158 frames in Fig. 5 where the instantaneous SNR is significantly lower than zero.

2.6. *Aurora2 Evaluation Results*

Noise-robust connected digit recognition results obtained using the best version of SPLICE are shown in Fig. 6 for the full Aurora2 evaluation test data. Sets-A and -B each consists of 1101 digit sequences for each of four noise conditions and for each of the 0 dB, 5 dB, 10 dB, 15 dB, and 20 dB SNRs. The same is for Set-C except there are only two noise conditions. All the results in Fig. 6 are obtained with the use of cepstral mean normalization (CMN) for all data after applying noise-normalized, dynamic SPLICE to cepstral enhancement. The use of CMN has substantially improved the recognition rate for Set-C. For simplicity purposes, we have assumed no channel distortion in the implementation of the iterative stochastic approximation algorithm for noise estimation. This assumption would not be appropriate for Set-C which contains unknown but fixed channel distortion. This deficiency has been, at least partially, offset by the use of CMN.

The word error rate reduction achieved as shown in Fig. 6 is 27.9 % for the multi-condition training mode, and 67.4% for the clean-only training mode,

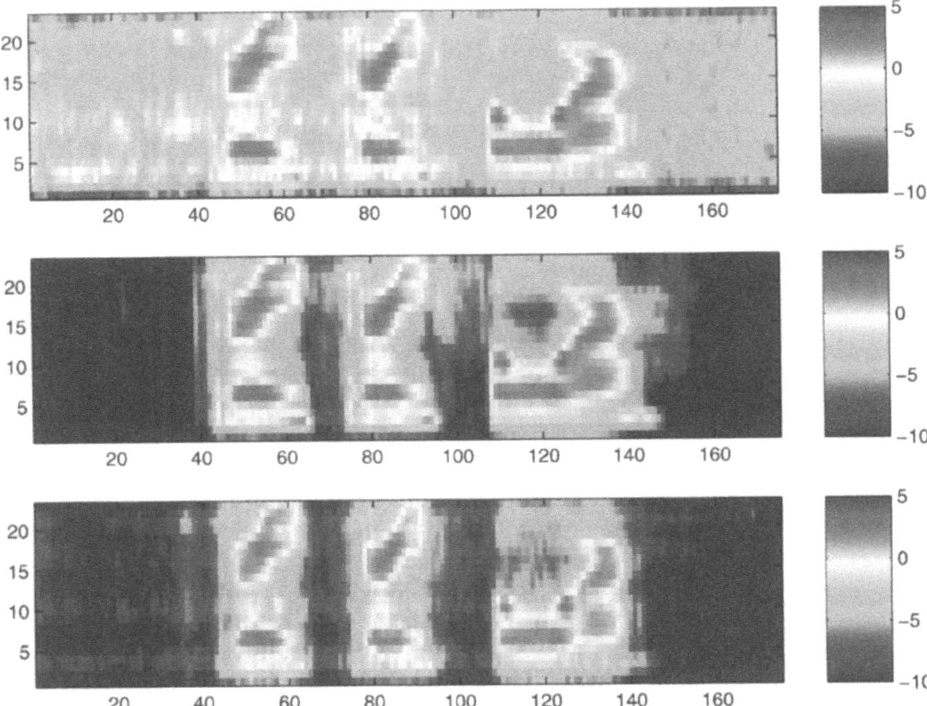

Figure 3. Noise-normalized SPLICE denoising using the iterative stochastic algorithm for tracking nonstationary noise in an utterance of the Aurora2 data with an average SNR = 10 dB. From top to bottom panels are noisy speech, clean speech, and denoised speech, all in the same spectrogram format.

respectively, compared with the results using the standard Mel cepstra with no speech enhancement. In the multi-condition training mode, the denoising algorithm is applied to the training data set and the resulting denoised Mel-cepstral features are used to train the HMMs. In the clean-training mode, the HMMs are trained using clean speech Mel-cepstra and the denoising algorithm is applied only to the test set. The results in Fig. 6 represent the best performance in the September-2001 AURORA2 evaluation. The experimental results also demonstrated the crucial importance of using the newly introduced iterations in improving the earlier stochastic approximation technique, and showed a varying degree of sensitivity, depending on the degree of noise nonstationarity, of the noise estimation algorithm's performance to the forgetting factor embedded in the algorithm [12].

3. Compression and Error Protection

In addition to noise robustness, we recently also started the work on the feature compression (source coding)

and error protection (channel coding) aspects of distributed speech recognition that is required by the client-server architecture for MiPad. This work is intended to address the three key requirements for successful deployment of distributed speech recognition associated with the client-server approach: (1) Compression of cepstral features (via quantization) must not degrade the speech recognition performance; (2) The algorithm for source and channel coding must be robust to packet losses, bursty losses or otherwise; and (3) The total time delay due to the coding, which results from a combined quantization delay, error-correction coding delay, and transmission delay, must be kept within an acceptable level. In this section, we outline the basic approach and preliminary results of this work.

3.1. Feature Compression

A new source coding algorithm has been developed that consists of two sequential stages. After the standard Mel- cepstra are extracted, each speech frame is first classified to a phonetic category (e.g., phoneme)

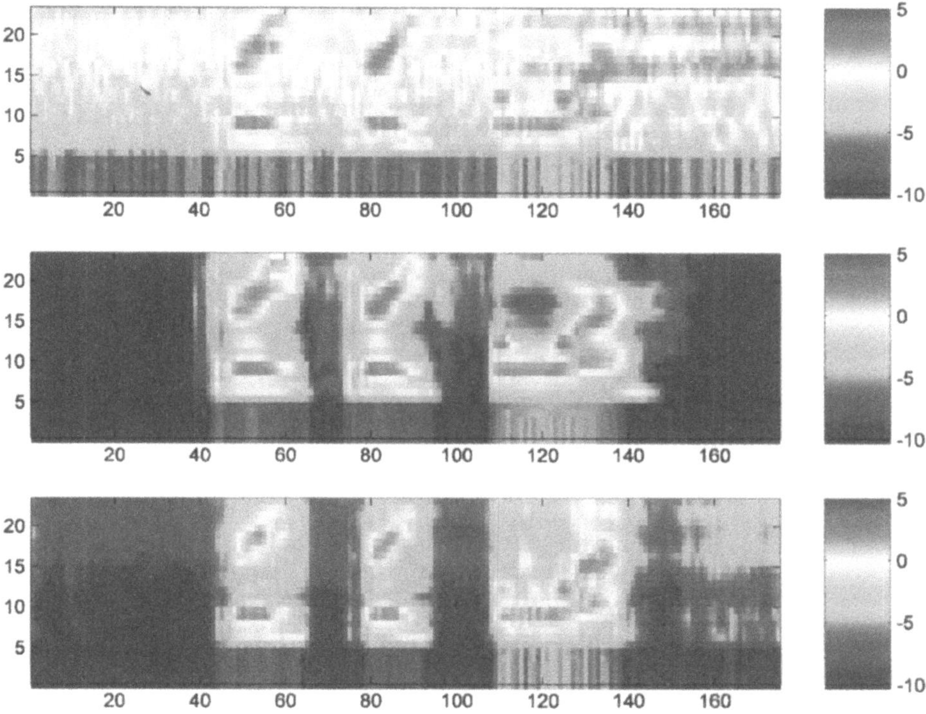

Figure 4. Noise-normalized SPLICE denoising using the iterative stochastic algorithm for tracking nonstationary noise in an utterance of the Aurora2 data with an average SNR = 5 dB. From top to bottom panels are noisy speech, clean speech, and denoised speech, all in the same spectrogram format.

and then is vector quantized (VQ) using the split-VQ approach. The motivation behind this new source coder is that the speech signal can be considered piecewise-stationary segments, and therefore can be most efficiently coded using one of many small codebooks that is tuned into that particular segment. Also, the purpose of the source coding considered here is to reduce the effect of coding on the speech recognizer's word error rate on the server-side of MiPad, which is very different from the usual goal of source coding aiming at maintaining perceptual quality of speech. Therefore, the use of phone-dependent codebooks is deemed most appropriate since phone distinction can be enhanced by using separate codebooks for distinct phones. Phone distinction often leads to word distinction, which is the goal of speech recognition and also the ultimate goal of feature compression designed for this purpose in the MiPad design.

One specific issue to be addressed in the coder design is bit allocation, or the number of bits that must be assigned to the subvector codebooks. In our coder, C0, C1–6, and C7–12 form three separate sets of subvec-

tors that are quantized independently (i.e., $M = 3$, where M is the total number of independent codebooks). Starting from 0 bit for each subvector codebook of each phone we can evaluate every possible combination of bits to subvectors and select the best according to a certain criterion. To better match the training procedure to the speech recognition task we use the criterion of minimal word error rate (WER). That is, big assignment is the result of following constrained optimization:

$$\hat{B} = \arg\min_{B}\{WER(B)\},$$

under the constraint of $\sum_i b_i = N$, where b_i is the number of bits assigned to the i-th subvector, N is the total number of bits to be assigned, and $WER(B)$ is the WER by using B as the assignment of the bits to subvectors. For the full search case, because for each one of the possible combinations we must run a separate WER experiment and the total number of combinations is prohibitively high we use a greedy bit allocation technique. At each stage we add a bit at each one of the

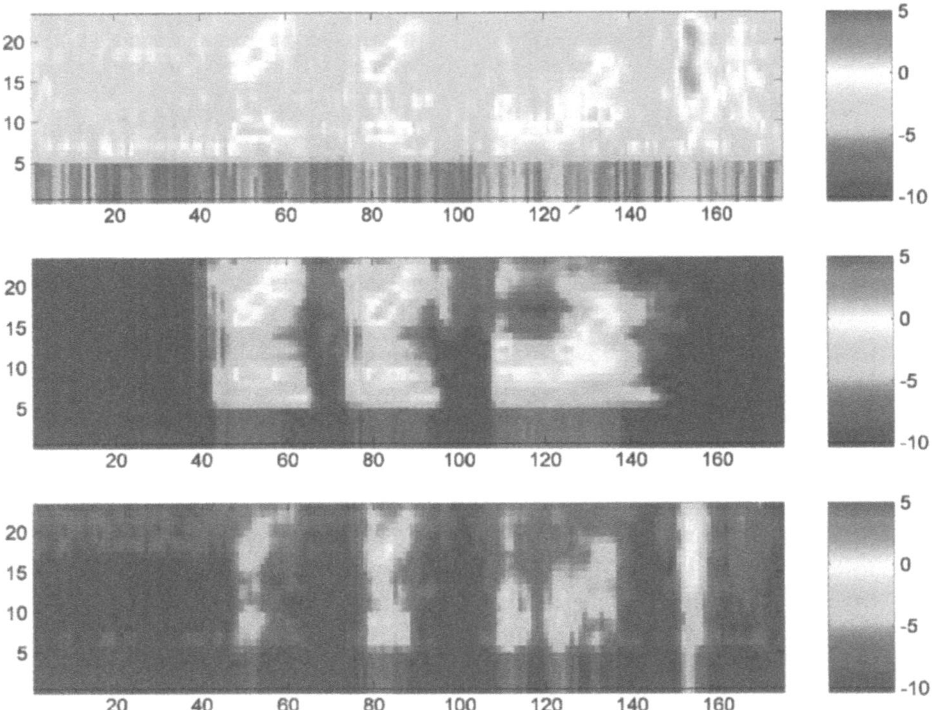

Figure 5. Noise-normalized SPLICE denoising using the iterative stochastic algorithm for tracking nonstationary noise in an utterance of the Aurora2 data with an average SNR $=0$ dB. From top to bottom panels are noisy speech, clean speech, and denoised speech, all in the same spectrogram format.

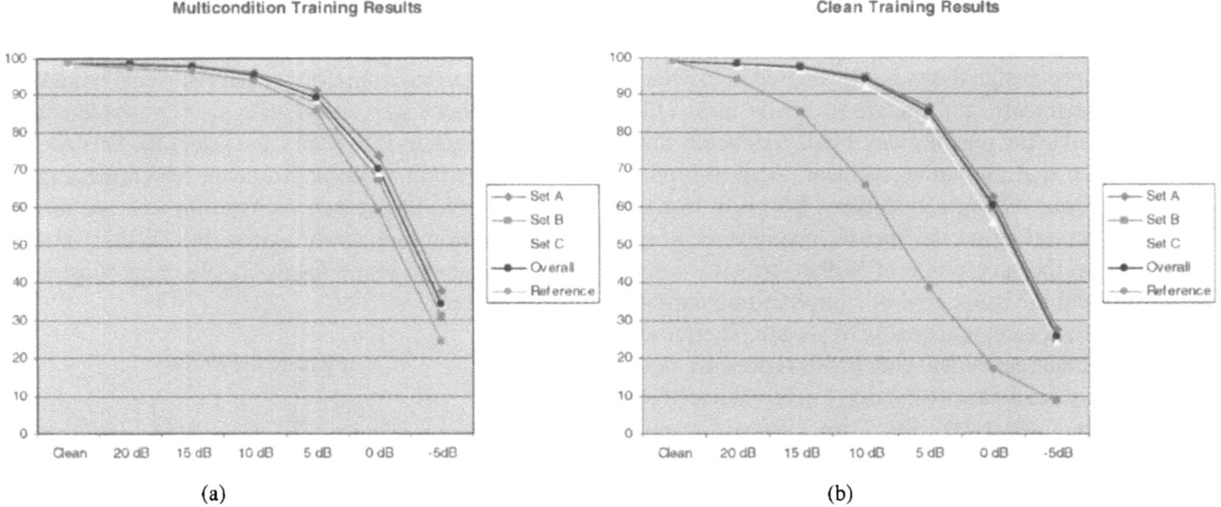

Figure 6. Full set of noise-robust speech recognition results in the September-2001 Aurora2 evaluation, using the dynamic and noise-normalized SPLICE with the noise estimation obtained from iterative stochastic approximation; Sets A, B, and C are separate test sets with different noise and channel distortion conditions. In (a) are the recognition rates using multi-condition training mode where the denoising algorithm is applied to the training data set and the resulting denoised Mel-cepstral features are used to train the HMMs. In (b) are the recognition rates using the "clean" training mode where the HMMs are trained using clean speech Mel-cepstra and the denoising algorithm is applied only to the test set. Reference curves in both (a) and (b) refer to the recognition rates obtained with no denoising processing.

subvectors and we keep the combination with the minimal WER. We repeat the procedure for the next stage by starting at the best combination of the previous step. By having M subvectors and N total bits to assign the total number of combinations is reduced from M^N to $M \times N$.

The experiments carried out to evaluate the above phone-dependent coder use the baseline system with a version of the Microsoft continuous-density HMMs (Whisper). The system uses 6000 tied HMM states (senones), 20 Gaussians per state, Mel-cepstrum, delta cepstrum, and delta-delta cepstrum. The recognition task is 5000-word vocabulary, continuous speech recognition from Wall Street Journal data sources. A fixed, bigram language model is used in all the experiments. The training set consists of a total of 16,000 female sentences, and the test set of 167 female sentences (2708 words). The word accuracy with no coding for this test set was 95.7%. With use of a perfect phone classifier, the coding using the bit allocation of $(4, 4, 4)$ for the three subvectors gives word accuracy of 95.6%. Using a very simple phone classifier with Mahalanobis distance measure, the recognition accuracy drops only to 95.0%. For this high-performance coder, the bandwidth has been reduced to 1.6 Kbps with the use of coder memory of 64 Kbytes.

3.2.   Error Protection

In the recent work, a novel channel coder has also been developed to protect the Mel-cepstral features for MiPad speech recognition based on the client-server architecture. The channel coder assigns unequal amounts of redundancy among the different source code bits, giving a greater amount of protection to the most important bits. The greater contributions the bits make to reducing the word error rate in speech recognition, the more important these bits are. A quantifiable procedure to assess the importance of each bit is developed, and the channel coder exploits this utility function for the optimal forward error correction (FEC) assignment. The FEC assignment algorithm assumes that packets are lost according to a Poisson process. Simulation experiments are performed where the bursty nature of loss patterns are taken into account. When combined with the new source coder, the new channel coder is shown to provide considerable robustness to packet losses even under extremely adverse conditions.

Some alternatives to FEC coding are also explored, including the use of multiple transmission, interleaving, and interpolation. We conclude from this preliminary work that the final choice of channel coder should depend on the relative importance among such factors as delay, bandwidth, and tolerance to the burstiness of noise.

Our preliminary work on the compression and error protection aspects of distributed speech recognition has provided clear insight into the tradeoffs we need to make between source coding, delay, computational complexity and resilience to packet loss. Most significantly, the new algorithms developed have been able to bring down the Mel-cepstra compression rate to as low as 1.6 Kbps with virtually no degradation in word error rate compared with no compression. These results are currently being incorporated into the next version of the MiPad design.

4.   Continuous Speech Recognition and Language Modeling

While the compressed and error-protected Mel-cepstral features are computed in the MiPad client, major computation for continuous speech recognition (decoding) resides in the host computer as the server. The entire set of the statistical language model, the acoustic model in the form of hidden Markov models (HMMs), and the lexicon that are used for speech decoding all reside in the server, processing the Mel-cepstral features transmitted from the client. Denoising operations such as SPLICE that extract noise-robust Mel-cepstra can reside either on the server or on the client, or on both working in collaboration.

The MiPad is designed to be a *personal* device. As a result, the recognition uses speaker-adaptive acoustic models (HMMs) and a user-adapted lexicon to improve recognition accuracy. The HMMs and the continuous speech decoding engine are both derived from an improved version of the Microsoft's Whisper speech recognition system and of the HTK, which combines the best features of these earlier two separate systems. Both MLLR (Maximum Likelihood Linear Regression) and MAP adaptation are used to adapt the speaker-independent acoustic model for each individual speaker. There are 6000 senones, each with 20-component mixture Gaussian densities. The context-sensitive language model is used for relevant semantic objects driven by the user's pen tapping action, as will be described in the MiPad's Tap and Talk interface design in Section 6. As speech recognition accuracy remains as a major challenge for MiPad usability, most of our recent work on MiPad's acoustic modeling has focused on noise robustness as described in Section 2. The work on language modeling for improving speech recognition accuracy has focused on language model portability, which is described in this section below.

The speech recognition engine in MiPad uses the unified language model [6] that takes advantage of both rule-based and data-driven approaches. Consider two training sentences:

- *"Meeting at three with Zhou Li". vs.*
- *"Meeting at four PM with Derek".*

Within a pure n-gram framework, we will need to estimate

$$P(Zhou \mid three\ with) \quad and \quad P(Derek \mid PM\ with)$$

individually. This makes it very difficult to capture the obviously needed long-span semantic information in the training data. To overcome this difficulty, the unified model uses a set of CFGs that captures some of the common named entities. For the example listed here, we may have CFG's for <NAME> and <TIME> respectively, which can be derived from the factoid grammars of smaller sizes. The training sentences now look like:

- *"Meeting <at three:TIME> with <Zhou Li:NAME>",* and
- *"Meeting <at four PM:TIME> with <Derek: NAME>".*

With parsed training data, we can now estimate the n-gram probabilities as usual. For example, the replacement of

$$P(Zhou \mid three\ with) \leftarrow P(<NAME> \mid\\ <TIME>\ with)$$

makes such "n-gram" representation more meaningful and more accurate.

Inside each CFG, however, we can still derive

$$P(\text{"Zhou Li"} \mid <NAME>) \quad and$$
$$P(\text{"four PM"} \mid <TIME>)$$

from the existing n-gram (n-gram probability inheritance) so that they are appropriately normalized [16]. This unified approach can be regarded as a generalized n-gram in which the vocabulary consists of words and structured classes. The structured class can be simple, such as <DATE>, <TIME>, and <NAME>, if there is no need to capture deep structural information. It can be made complicated also in order to contain deep structured information. The key advantage of the unified language model is that we can author limited CFGs for each new domain and embed them into the domain-independent n-grams. In short, CFGs capture domain-specific structural information that facilitates language model portability, while the use of n-grams makes the speech decoding system robust against catastrophic errors.

Most decoders can only support either CFGs or word n-grams. These two ways of representing sentence probabilities were mutually exclusive. We have modified the decoder so that we can embed CFGs in the n-gram search framework to take advantage of the unified language model. An evaluation of the use of the unified language model is shown in Table 1. The speech recognition error rate with the use of the unified language model is demonstrated to be significantly lower than that with the use of the domain-independent trigram. That is, incorporating the CFG into the language model drastically improves cross-domain portability. The test data shown in Table 1 are based on MiPad's PIM *conversational speech*. The domain-independent trigram language model is based on Microsoft Dictation trigram models used in Microsoft Speech SDK 4.0. In Table 1, we also observe that using the unified language model directly in the decoding stage produces

Table 1. Cross-domain speaker-independent speech recognition performance with the unified language model and its corresponding decoder.

Systems	Perplexity	Word error (%)	Relative decoding time
Domain-independent trigram	593	35.6	1.0
Unified decoder with the unified LM	141	22.5	0.77
N-best re-scoring with the unified LM	–	24.2	–

about 10% fewer recognition errors than doing N-best re-scoring using the identical language model. This demonstrates the importance of using the unified model in the early stage of speech decoding.

5. Spoken Language Understanding and Dialogue Management

The spoken language understanding (SLU) engine used in our speech-centric multimodal human-computer interaction research, in MiPad research in particular, is based on a robust chart parser [7] and a plan-based dialog manager [13]. Each semantic object defined and used for SLU is either associated with a real-world entity or an action that the application takes on a real-entity. Each semantic object has slots that are linked with their corresponding CFG. In contrast to the sophisticated prompting response in voice-only conversational interface, the response is a direct graphic rendering of the semantic object on MiPad's display. After a semantic object is updated, the dialog manager fulfills the plan by executing application logic and error repair strategy.

One of the critical tasks in SLU is semantic grammar authoring. Manual development of domain-specific grammars is time-consuming, error-prone, and it requires a significant amount of expertise. We have been working on semi-automatic grammar learning tools that take advantage of multiple information sources to help developers author domain-specific semantic grammars.

In this section, we will describe in detail our recent research activities outlined above.

5.1. Semantic Schema and Knowledge Representation

MiPad adopts a semantic based robust understanding technology for spoken language understanding. At the center of the technology is *semantic schema* defined in the Semantic Description Language (SDL). The semantic schema is a domain model; it defines the entity relations of a specific domain. The semantic schema is used for many different purposes. It serves as the specification for a language-enabled application: once a semantic schema is defined, grammar and application logic development can proceed simultaneously according to the semantic schema. It also plays a critical role in dialogue management. Further, semantic schema is

language and expression independent in the sense that it does not specify the linguistic expressions used to express the concepts. Because of this, it is used not only for language-enabled applications, but also for integrating inputs from multi-modalities, such as mouse click events. Below is an example of concept definitions in a semantic schema:

```
<command type="AppointmentUpdate" name=
   "AddAttendee">
   <slot type="People"/>
   <slot type="ExistingAppt"/>
</command>
<command type="AppointmentUpdate" name=
   "ScheduleAppointment">
   <slot type="People"/>
   <slot type="Time" name="StartTime"/>
   <slot type="Time" name="EndTime"/>
</command>
<entity type="ExistingAppt" name=
   "ApptByAttributes">
   <slot type="People"/>
   <slot type="Time" name="StartTime"/>
   <slot type="Time" name="EndTime"/>
</entity>
<entity type="People" name="ByName">
   <slot class="FirstName"/>
   <slot class="LastName"/>
</entity>
<verbatim type="Name" name="Firstname"/>
<verbatim type="Name" name="Lastname"/>
```

A semantic schema consists of a list of definitions for semantic classes. A semantic class corresponds to a concept in the application domain. The example above shows three different kinds of semantic classes: *command*, *entity* and *verbatim*. Command and entity are semantic classes that contain component slots, while verbatim is a semantic terminal. Each semantic class has a type, and multiple semantic classes may share the same type. For example, the type "People" can be shared by semantic classes "ByName" (e.g., Peter Johnson), "ByReportingRelation" (e.g., my manager), "ByAnaphora" (e.g. him), or "ByClick" (e.g. Mouse Clicking on a person's picture). Slots of a semantic class are specified with either a type or a semantic class. The former constrains that the slot must be filled with a semantic object (an instantiation of a semantic class) that has the specified type, and the latter requires that the slot be filled with an instantiation of that specific semantic class. In case two slots are specified with the same type or semantic class, additional names are used to differentiate them (e.g., StartTime and EndTime in the above example.)

In a human-machine conversation, the computer system responds to the semantics (denoted by S) of a user's utterance (word sequence w) with an appropriate action (A). It does so with the help of the discourse structure (D), which accumulates over all the relevant semantic information from the beginning of the conversation up to the current utterance. Both the utterance semantics and the discourse information are represented in an XML structure that maps words in the utterances to the semantic classes (including all their slots). The following is a concrete example of a dialogue, together with the utterance semantics and discourse structures after each of the user's turns in the dialogue.

w1:   Schedule a meeting with Peter.
A1:   Peter who?
w2:   Peter Johnson.
A2:   When do you want to start the meeting?
w3:   Tuesday at 2 pm.

After the user utters w1, its meaning can be represented as the following (partial) semantic object:

S1:   <ScheduleAppointment type =
          "AppointmentUpdate">
          <ByName type= "Person">
              <FirstName type= "Name"> Peter
                  </FirstName>
          </ByName>
      </ScheduleAppointment>

Since this is the first utterance of the dialogue, the discourse structure D1 = S1.

By comparing D1 with the semantic class definition of ByName, the system knows that it cannot resolve the ambiguity without the last name for Peter. Therefore, it takes the appropriate action (A1) to ask the user for the last name of Peter. After the user responds with the reply w2, the SLU system "understands" w2 by generating the following utterance semantics (via robust parsing):

S2:   <ByName type= "Person">
          <FirstName type= "Name"> Peter
              </FirstName>
          <LastName type= "Name"> Johnson
              </LastName>
      </ByName>

And the new discourse structure is obtained by adjoining S2 to D1:

D2:   <ScheduleAppointment type=
          "AppointmentUpdate">
          <ByName type="People">
              <FirstName type= "Name"> Peter
                  </FirstName>
              <LastName type= "Name"> Johnson
                  </LastName>
          </ByName>
      </ScheduleAppointment>

After w2, the system resolves the ByName semantic object. It then tries to resolve the parent semantic object "ScheduleAppointment". By comparing D2 with the semantic class definition for "ScheduleAppointment", it knows that the time information for the appointment is missing. Therefore, it prompts the user with A2. After the user replies with w3, it generates the following utterance semantic object and augments the discourse structure accordingly:

S3:   <Time type = "Time" name = "StartTime">
          Tuesday 2 pm </Time>

D3:   <ScheduleAppointment type=
          "AppointmentUpdate">
          <ByName type= "People">
              <FirstName type= "Name"> Peter
                  </FirstName>
              <LastName type= "Name"> Johnson
                  </LastName>
          </ByName>
          <Time type = "Time" name =
              "StartTime"> Tuesday 2 pm </Time>
      </ScheduleAppointment>

After the system resolves all ambiguities about the meeting by completing all the slots in the semantic class specified in the schema, it will take the appropriate action to set the meeting in the calendar and inform the attendees about it.

The example above illustrates the underlying working principle of the language processing component in the MiPad design. In the next section, we will provide a more rigorous mathematical framework for such a component.

5.2. A Unified Framework for Speech Recognition, Understanding, and Dialogue Management

The SLU and dialogue management components in our multimodal interaction system's architecture exemplified by MiPad can be integrated with the speech recognition component, and, together, be placed into a unified pattern recognition framework employing powerful optimization techniques in system engineering. Using the same notation as the previous example, we let S_n and D_n denote the utterance semantic object and discourse structure, respectively, and A_n the system's action, all after the n-th dialog turn. Given a new acoustic signal of speech $\mathbf{y}$, the SLU problem can be formulated as the following optimization one:

$$\begin{aligned}
\hat{D}_n &= \arg\max_{D_n} P\left(D_n \mid \mathbf{y}, D_{n-1}, A_1^{n-1}\right) \\
&= \arg\max_{D_n} \sum_w P\left(D_n \mid w, \mathbf{y}, D_{n-1}, A_1^{n-1}\right) \\
&\quad \times P\left(w \mid \mathbf{y}, D_{n-1}, A_1^{n-1}\right),
\end{aligned} \qquad (10)$$

where w is the word sequence (in the current dialogue turn) corresponding to the speech utterance $\mathbf{y}$, and is marginalized out on the right hand side of Eq. (10) using the weighting function $P(w \mid \mathbf{y}, D_{n-1}, A_1^{n-1})$ provided by the speech recognizer. The speech recognizer computes this posterior probability according to

$$\begin{aligned}
&P\left(w \mid \mathbf{y}, D_{n-1}, A_1^{n-1}\right) \\
&= \frac{P\left(\mathbf{y} \mid w, D_{n-1}, A_1^{n-1}\right) P\left(w \mid D_{n-1}, A_1^{n-1}\right)}{\sum_w P\left(\mathbf{y} \mid w, D_{n-1}, A_1^{n-1}\right) P\left(w \mid D_{n-1}, A_1^{n-1}\right)} \\
&\approx \frac{P(\mathbf{y} \mid w) P(w \mid D_{n-1}, A_{n-1})}{\sum_w P(\mathbf{y} \mid w) P(w \mid D_{n-1}, A_{n-1})},
\end{aligned} \qquad (11)$$

where $P(\mathbf{y} \mid w, D_{n-1}, A_1^{n-1}) = P(\mathbf{y} \mid w)$ is the acoustic score from the recognizer,[1] and $P(w \mid D_{n-1}, A_1^{n-1}) = P(w \mid D_{n-1}, A_{n-1})$ is the dialogue-state dependent language model score. For the discourse model $P(D_n \mid y, w, D_{n-1}, A_1^{n-1})$, since D_n can be deterministically obtained from D_{n-1} and S_n, as their adjoin, we can simplify it to

$$P\left(D_n \mid \mathbf{y}, w, D_{n-1}, A_1^{n-1}\right) = P\left(S_n \mid \mathbf{y}, w, D_{n-1}, A_1^{n-1}\right).$$

In practice, we further assume conditional independence between the speech acoustics $\mathbf{y}$ and the semantic object S_n given word sequence w, and make a Markov assumption on the action sequence A_1^{n-1}. This thus further simplifies the discourse model component in Eq. (10) to

$$P\left(S_n \mid \mathbf{y}, w, D_{n-1}, A_1^{n-1}\right) = P(S_n \mid w, D_{n-1}, A_{n-1}).$$

We now take into account the acoustic denoising operation that nonlinearly transforms the generally distorted speech signal $\mathbf{y}$ into its estimated undistorted version $\hat{\mathbf{x}}$. Using the example of the SPLICE algorithm as described in Section 2, we have the relationship between the distorted and estimated undistorted speech features:

$$\hat{\mathbf{x}} = g(\mathbf{y}) = \mathbf{y} + \sum_s p(s \mid \mathbf{y}) \mathbf{r}_s(\mathbf{y}).$$

Further, we use Viterbi approximation to reduce the computational load in the overall optimization operation shown in the framework of Eq. (10) for the SLU problem. Taking account all the above considerations and approximations, the SLU problem formulated in Eq. (10) becomes drastically simplified to

$$\begin{aligned}
\hat{S}_n &\approx \arg\max_{S_n, w} P(S_n \mid w, D_{n-1}, A_{n-1}) \\
&\quad \times P(w \mid \hat{\mathbf{x}}, D_{n-1}, A_{n-1}) \\
&= \arg\max_{S_n, w} P(S_n \mid w, D_{n-1}, A_{n-1}) \\
&\quad \times \frac{P(\hat{\mathbf{x}} \mid w, D_{n-1}, A_{n-1}) P(w \mid D_{n-1}, A_{n-1})}{\sum_w P(\hat{\mathbf{x}} \mid w, D_{n-1}, A_{n-1}) P(w \mid D_{n-1}, A_{n-1})}
\end{aligned}$$
$$(12)$$

We note that even with all these simplifications, significant challenges remain due to the data sparseness problem in training the semantic model $P(S_n \mid w, D_{n-1}, A_{n-1})$ and due to the large computational requirement for evaluating the denominator in the above equation.

Based on the formulation of the SLU solution discussed above, which seeks the maximal probability $P(S_n \mid \hat{\mathbf{x}}, D_{n-1}, A_{n-1})$, or equivalently the maximal product of the two probabilities: $P(S_n \mid w, D_{n-1}, A_{n-1}) P(w \mid \hat{\mathbf{x}}, D_{n-1}, A_{n-1})$, the dialog management problem can also be formulated as an optimization problem as follows: For the n-th turn, find an action A such that the averaged cost function $C(A, D_n)$ is minimized over the conditional probability measure of $P(S_n \mid \hat{\mathbf{x}}, D_{n-1}, A_{n-1})$. That is,

$$\begin{aligned}
\hat{A} &= \arg\max_A E[C(A, D_n) \mid \hat{\mathbf{x}}, D_{n-1}, A_{n-1}] \\
&= \arg\max_A \sum_{D_n} C(A, D_n) P(S_n \mid \hat{\mathbf{x}}, D_{n-1}, A_{n-1}).
\end{aligned}$$

The principal challenge in implementing this dialog manager is again the data sparseness problem in determining the cost function $C(A, D_n)$ and the discourse model $P(S_n \mid w, D_{n-1}, A_{n-1})$. In our current implementation for multimodal interaction such as MiPad, the data sparseness problem in both SLU and dialog management is partially by-passed using empirical rules in approximating the probabilities formulated above. One key element in this implementation is the (deterministic) semantic grammar-based robust parser that is used to map from w to S_n with the help of the dialogue state (D_{n-1}, A_{n-1}). The grammar will need to be generalized to a probabilistic one in order to compute the necessary probability terms in the unified framework discussed above. We now describe our recent research on the semantic based robust understanding that is in the current implementation of the SLU system in MiPad design.

5.3. Robust Parser and SLU in MiPad

Since the *Tap & Talk* interface in MiPad explicitly provides dialog state (tapped field) information already, dialogue management plays relatively minor role, compared with the SLU, in the overall MiPad functionality. The major SLU component in MiPad is a robust chart parser, which accepts the output of the continuous speech recognizer using field-specific language models and employs field-specific grammars. In the typical MiPad usage scenario, users use the built-in MiPad microphone that is very sensitive to environment noise. With the iPaq device from Compaq as one of our prototypes, the word recognition error rate increased by a factor of two in comparison to a close-talking microphone in the normal office environment. This highlights the need not only for noise-robust speech recognition but also for robust SLU.

The MiPad SLU is modeled with domain-specific semantic grammars. Normally, semantic grammars are CFGs with non-terminals representing semantic concepts instead of syntactic categories. Our grammar introduces a specific type of non-terminals called semantic classes to describe the schema of an application. The semantic classes define the conceptual structures of the application that are independent of linguistic structures. The linguistic structures are modeled with context free grammars. In doing so, it makes the linguistic realization of semantic concepts transparent to an application; therefore the application logic can be implemented according to the semantic class structure, in parallel with the development of linguistic context free grammar. We in the past few years have developed a robust spoken language parser that analyzes input sentences according to the linguistic grammar and maps the linguistic structure to the semantic conceptual structure. Recently, we have made substantial modifications to the parser to take full advantage of the form factor of MiPad and to better support the semantic based analysis.

5.3.1. Robust Chart Parser. The robust parsing algorithm used in MiPad is an extension of the bottom-up chart-parsing algorithm. The robustness to ungrammaticality and noise can be attributed to its ability of skipping minimum unparsable segments in the input. The algorithm uses dotted rules, which are standard CFG rules plus a dot in front of a right-hand-side symbol. The dot separates the symbols that already have matched with the input words from the symbols that are yet to be matched. Each constituent constructed in the parsing process is associated with a dotted rule. If the dot appears at the end of a rule like in $\mathbf{A} \rightarrow \alpha\bullet$, we call it a complete parse with symbol $\mathbf{A}$. If the dot appears in the middle of a rule like in $\mathbf{A} \rightarrow \mathbf{B}\bullet\mathbf{CD}$, we call it a partial parse (or hypothesis) for A that is expecting a complete parse with root symbol $\mathbf{C}$.

The algorithm maintains two major data structures—A chart holds hypotheses that are expecting a complete constituent parse to finish the application of the CFG rules associated with those hypotheses; an agenda holds the complete constituent parses that are yet to be used to expand the hypotheses in the chart. Initially the agenda is empty. When the agenda is empty, the parser takes a word (from left to right) from the input and puts it into the agenda. It then takes a constituent $\mathbf{A}[i, j]$ from the agenda, where A is the root symbol of the constituent and $[i, j]$ specifies the span of the constituent. The order by which the constituents are taken out of the agenda was discussed in [12]. The parser then activates applicable rules and extends appropriate partial parses in the chart. A rule is applicable with respect to a symbol $\mathbf{A}$ if either A starts the rule or all symbols before $\mathbf{A}$ are marked optional. The activation of an applicable rule may result in multiple constituents that have the same root symbol (the left-hand-side of the rule) but different dot positions, reflecting the skip of different number of optional rule symbols after $\mathbf{A}$. If the resulting constituent is a complete parse, namely with the dot positioned at the end of the rule, the complete constituent is added into the agenda. Otherwise partial constituents are added into the chart; To extend the partial parses with the complete parse $A[i, j]$, the parser exams the chart for incomplete constituent with dotted rule $B[l, k] \rightarrow \alpha \bullet A\beta$ for $k < i$, and constructs new constituents $B[l, j] \rightarrow \alpha\, A\beta$ with various dot positions in β, as long as all the symbols between A and the new dot position are optional. The complete constituent $B[l, j] \rightarrow \alpha\, A\beta\bullet$ is added into the agenda. Other constituents are put into the chart. The parser continues the above procedure until the agenda is empty and there are no more words in the input sentence. By then it outputs top complete constituents according to some heuristic scores.

In [7], we distinguished three different types of rule symbols: optional symbols that can be skipped without penalty; normal symbols that can be skipped with penalty; and mandatory symbols that cannot be skipped. We found the skip of normal symbols is very problematic, because grammar authors are generally very forgetful to mark a symbol mandatory. This is also because skipping normal rule symbols often adversely increases the constituent space. This dramatically slows down the parser and results in a great number of bogus ambiguities. Therefore, in our current parser implementation, we do not skip rule symbols unless the symbols are explicitly marked as optional.

5.3.2. Special Features of the Robust Parser.

To take advantage of the MiPad form factor and better support semantic analysis, we have enhanced the parser with the four new features that are described below.

5.3.2.1. Dynamic Grammar Support.

Dynamic grammar support provides the parser with the capability of modifying the grammar it is using on the fly. It is necessary because different users may have different data; therefore the parser should be able to customize the grammar online. For example, users may have different and changing contact list, therefore the parser should dynamically modify the rule for the contacts of different users. Dynamic grammar support is necessary also because different dialog states need different grammars too. If MiPad is showing a New-Email card, then the ByName semantic class should only contain those names in the user's contact list, since they are the only names that the user can specify as recipients; otherwise the user has to specify an e-mail address. On the other hand, if MiPad is showing a New-Contact card, then we should use a name set with a much greater coverage, perhaps even introducing a spelling grammar.

We have devised an API for application developers to dynamically and efficiently change the grammar used by the parser. The change can be made at different granularity, from the entire grammar to every single rule. This enables the parser to adapt to different users and dialog states as appropriate.

5.3.2.2. Wildcard Parsing.

In a semantic grammar, some semantic concepts are free-form texts and can hardly be modeled with semantic rules. For instance, meeting subjects can hardly be predicated and modeled with semantic grammar rules. To model this type of semantic units, the parser is augmented to handle rules with wildcards like the following one:

```
<Meeting-Property> ::= <about>
   <Subject:Wildcard>
<about> ::= about | regarding | to discuss
```

Here "<Subject:Wildcard>" represents the non-terminal semantic class "<Subject>" that can match a free-form text. "<about>" serves as a context cue that triggers the wildcard rule for "<Subject>". Anything that does not match other appropriate rules in the context and that matches the wildcard rules with an appropriate trigger will be parsed as a wildcard semantic component in the appropriate context.

5.3.2.3. Parsing with Focus.

The parser approximates the statistical discourse model $P(S_n \mid w, D_{n-1}, A_{n-1})$ by taking advantage of the dialogue state information (D_{n-1}, A_{n-1}) to reduce parsing ambiguities and the search space, hence to improve its performance. For example, if the parser knows that the system is in the middle of a dialog with the user, talking about the attendee of a meeting, then the parser will only parse "John Doe" as Attendee, although it could be E-mail Recipient or new Contact according to the grammar. The parser can get the context information (focus) either from the dialog manager in the general *Dr. Who* architecture or directly from the applications with the *Tap & Talk* interface. The focus is specified as a path from the root to the semantic class that the system is expecting an input for, such as Root/ScheduleMeeting/MeetingProperty/StartTime.

The parser can override the focus mode in case the user volunteers more information in mixed initiative dialogs. In this case, if the system is expecting an input of the aforementioned focus and the user speaks "from 3 to 5 pm", the parser will be smart enough to identify both Start-Time and End-Time, and return the semantic class that is the closest common ancestor of the identified semantic classes, which in this case is Meeting-Property.

5.3.2.4. N-Best Parsing.
The parser can take n-best hypotheses from the speech recognizer, together with their scores, as the input. The parser analyzes all the n-best hypotheses and ranks them with a heuristic score that combines the speech recognition score and the parsing score. The best parse will be forwarded to the application.

5.3.3. Practical Issues in the Parser Implementation.
In the parser implementation, it has been discovered that the parser often slowed down with the support of wildcard parsing. To overcome this weakness, we have redesigned the data structure, which leads to dramatic improvement in the parser efficiency. We briefly describe this and some related implementation issues below.

5.3.3.1. Lexicon Representation.
The lexicon contains both terminals and non-terminals. The non-terminals include names for semantic class, groups and productions. Each entry in the lexicon is represented with an ID. The lexicon can map a terminal or non-terminal string to its ID. An ID is an integer with the least signification 3 bits devoted to the type of the ID (word, semantic classes, types, productions, groups etc.) The rest bits can be used as index for the direct access to the definition of the grammar components of the specific type. Each lexical entry points to a set Φ that contains the information of the IDs of the context free rules it can activate, as well as the position of the lexical item in these rules. Set Φ is used to locate the applicable rules after a complete is taken out of the agenda.

Each non-terminal entry A has a Boolean element, specifying if it can derive a wildcard as its left-most descendant in a parse, or more formally, if $A \to$ wildcard α. The value of this element has to be pre-computed at grammar load time, and recomputed every time after dynamic grammar modification. The speed to set this Boolean value is hence very important. Fortunately we already have the information of rules that can be activated by a symbol in set Φ. With that information, we can define the relation $\Re = \{(x, b) \mid b \Rightarrow \alpha x \beta\}$, where α is empty or a sequence of optional symbols. Then a non-terminal can derive a wildcard on its leftmost branch if and only if it is in the transitive closure of the wildcard with respect to relation $\Re$. This transitive closure can be computed in time linear to the number of non-terminal symbols.

5.3.3.2. Chart and Agenda.
The chart consists of a dynamic array of n elements and a dynamic programming structure of $n^*(n + 1)/2$ (n is the length of an input sentence) cells that corresponds to the $n^*(n + 1)/2$ different span of constituents. Each array element corresponds to a position in the input sentence, and it contains a heap that maps from a symbol **A** to a list of partial parses. The partial parses cover the input sentence to the position that the element represents for, and they expect a complete parse of a constituent with root symbol **A**. With this the parser can quickly find the partial parses to extend when a complete parse with root **A** is popped from the agenda; Each cell of the dynamic programming structure contains a heap that maps a grammar symbol **A** to a pointer to the complete constituent tree with root **A** and the span that the cell represents for. This enables the parser to quickly find out if a new constituent has the same root name and span as an existing constituent. If so, the parser will safely prune the constituent with lower score.

The agenda is implemented as a priority queue. An element of the queue has a higher priority if it has a smaller span and higher heuristic score. This guarantees that the parser does not miss any parses with high heuristic score.

5.3.3.3. Wildcard Support.
Since wildcard match is expensive, we would like to treat input words as wildcard only when it fits in the context. Therefore we added some top down guidance for the creation of a wildcard constituent. With the wildcard derivation information available for non-terminal symbols as described in 0, this can be implemented efficiently: during the parsing process, after a partial constituent with dotted rule $A \to \alpha \bullet B\beta$ is added to the chart, if **B** can derive a wildcard on its leftmost branch, we then set a flag that allows the next input word to be introduced as a wildcard.

After we introduce a wildcard to the parser, theoretically it can build m different wildcard constituents with different coverage, where m is the number of remaining words in input. This adversely increases the search space drastically, since each of these wildcard constituents can be combined with other constituents to form much more constituents. Instead of generating these m constituents, we assume that the wildcard only covers a single word. After the parser has built the parse for the complete sentence, it expands the wildcard coverage to all the skipped words adjacent to the word covered by a wildcard in the initial parse. The parser always prefers non-wildcard coverage to wildcard coverage. So wildcard will be used only when there is no no-wildcard parse of the input segment that fits the context.

5.4. Machine-Aided Grammar Learning and Development for SLU

The SLU component implemented in MiPad is a variation of the general semantic-based robust understanding technology. The technology has been widely used in human/machine and human/human conversational systems. For example, many research labs have used it in the DARPA-sponsored Airline Travel Information System (ATIS) evaluations. Such implementations have relied on manual development of a domain-specific grammar, a task that is time-consuming, error-prone and requires a significant amount of expertise. If conversational systems are to be a mainstream, it becomes apparent that writing domain-specific grammars is a major obstacle for a typical application developer. Recently researchers have been working on tools for rapid development of mixed-initiative systems, but without addressing the problem of grammar authoring per se. Also, other researchers have developed tools that let an end user refine an existing grammar, which still relies on an initial grammar and also assumes that the developer has a good knowledge of language structures.

On the other hand, automatic grammar inference has also attracted the attention of researchers for many years, though most of that work has focused on toy problems. Applications of such approaches on grammar structure learning for natural language have not been satisfactory for natural language understanding application. This has been due to the complexity of the problem. That is, the available data will typically be sparse relative to the complexity of the target grammar, and there has not been a good generalization mechanism developed to correctly cover a large variety of language constructions.

Therefore, instead of aiming at an ambitious empirical automatic grammar inference, we focus in our research on an engineering approach that could greatly ease grammar development by taking advantage of many different sources of prior information. In doing so, a good quality semantic grammar can be derived semi-automatically with a small amount of data.

5.4.1. Multiple Information Sources Used for Constructing Semantic Grammar.
A semantic CFG, like a syntactic CFG, defines the legal combination of individual words into constituents and constituents into sentences. In addition, it also has to define the concepts and their relations in a specific domain. It is this additional dimension of variation that makes it necessary to develop a grammar for every new domain. While it is not realistic to empirically learn structures from a large corpus due to technical and resource constraints, we can greatly facilitate grammar development by integrating different information sources to semi-automatically induce language structures. These various information sources are described below.

5.4.1.1. Domain-Specific Semantic Information.
As we mentioned earlier at the beginning of this section, we use the semantic schema to define the entity relations of a specific domain in our multi-modal human-computer interaction research. The domain specific semantic information in the schema can be incorporated into a CFG to model the semantic constraints. Since the semantic schema is used for many different purposes, from dialogue modeling to multi-modal input integration, it has to be developed in the first place in a multi-modal application; therefore it is not an extra burden to use it in grammar learning. Due to its language- or expression-independency, semantic schema is easy to author by a developer with good knowledge of an application. For the MiPad's calendar domain described in [5], the schema contains 16 concepts (semantic object constituents) with fewer than 60 slots. This is two orders of magnitude lower than the $\sim$3000 CFG rules for $\sim$1000 nonterminals. Such a semantic schema can be easily developed within a couple of hours.

5.4.1.2. Grammar Library.
Some low level semantic entities, such as date, time, duration, postal address, currency, numbers, percentage, etc, are not domain-specific. They are isolated universal building blocks that can be written once and then shared by many applications. Grammar libraries can greatly save development time, and we have used them extensively in developing MiPad grammars.

5.4.1.3. Annotation.
We can also get developers involved to annotate the data against the schema. For

example, the sentence "invite Ed to the meeting with Alex" is annotated against the schema as follows:

```
<AddAttendee text="invite Ed to the meeting with
   Alex">
      <ApptByAttributes text="the meeting with
         Alex">
            <People text="Alex"/>
      </ApptByAttributes>
      <People text="Ed"/>
</AddAttendee>
```

Here, each XML tag is the name of a semantic class in the schema, and the sub-structures are the members of semantic objects that fill in the slots of a semantic class. The annotation is surface-structure (i.e., linguistic expression) independent—different sentences that convey the same meaning would have the same annotation. The use of the grammar library also eases the annotation process since we do not have to annotate to the very bottom level of concepts.

5.4.1.4. Syntactic Constraints. The final source of information for semantic grammar development is the syntactic constraints, as domain specific language must follow the syntactic constraints of the language. Some simple syntactic clues, for example, part-of-speech constraints, are used to reduce the search space in grammar development.

5.4.2. Growing Semantic Grammar

5.4.2.1. Inherit Semantic Constraints from Schema. An assumption we made is that the linguistic constraints that guide the integration of smaller units into a larger chunk is an invariant for the subset of the natural language used in human-computer interaction. It is only the domain-specific semantics and linguistic expressions for concepts that can vary. This allows us to create a template CFG that inherits the semantic constraints from the semantic schema. For example, the two concepts in the previous example can be automatically translated to the following template CFG:

```
<T_ExistingAppt> → <C_ApptByAttributes>                                                          (1)
<C_ApptByAttributes> → {<ApptByAttributeMods>} <ApptByAttributeHead>
   {<ApptByAttributeProperties>}                                                                 (2)
<ApptByAttributeProperties> → <ApptByAttributeProperty> {<ApptByAttributeProperties>}           (3)
<ApptByAttributeProperty> → <ApptByAttributePeopleProperty> |
                            <ApptByAttributeStartTimeProperty> |
                            <ApptByAttributeEndTimeProperty>                                     (4)
<ApptByAttributePeopleProperty> → {<PreApptByAttributePeopleProperty>} <T_People>
   {<PostApptByAttributePeopleProperty>}                                                         (5)
<ApptByAttributeHead> → NN                                                                       (6)
<PreApptByAttributePeopleProperty> →.* <T_UpdateAppt> → <C_AddAttendee>
<C_AddAttendee> → <AddAttendeeCmd> {<AddAttendeeProperties>}
<AddAttendeeCmd> →.*
<AddAttendeeProperties> → <AddAttendeeProperty> {<AddAttendeeProperties>}
<AddAttendeeProperty> → <AddAttendeePeopleProperty> | <AddAttendeeExistingApptProperty>
<AddAttendeeExistingApptProperty>→ {<PreAddAttendeeExistingApptProperty>} <T_ExistingAppt>
   {<PostAddAttendeeExistingApptProperty>
<AddAttendeePeopleProperty> → {<PreAddAttendeePeopletProperty>} <T_People>
   {<PostAddAttendeePeopletProperty> }
<PreAddAttendeeExistingApptProperty →.*
<PostAddAttendeeExistingApptProperty →.*
<PreAddAttendeePeopleProperty →.*
<PostAddAttendeePeopleProperty →.*
```

Here an *entity*, like ApptByAttributes, consists of a head, optional (in braces) modifiers that appear in front of the head (e.g. "*Alex's* meeting"), and optional properties that follow the head (e.g. "meeting *with Alex*") (rule 2). Both modifiers and properties are defined recursively, so that they finally incorporate a sequence of different slots (rules 3–4). Each slot

5.4.2.2. Annotation: Reducing the Search Space. The annotation reduces the search space for the rewriting rules for the pre-terminals: the annotated slots serve as the divider that localizes the learning space. For example, with the semantic annotation just described and the template CFG, our robust parser can obtain the partial parse shown below:

```
AddAttendee
     AddAttendeeCmd invite
     AddAttendeeProperties
          AddAttendeeProperty
               AddAttendeePeopleProperty
                    T_People Ed
                    PostAddAttendeePeopleProperty
          AddAttendeeProperties
               AddAttendeeProperty
                    AddAttendeeExistingApptProperty
                         PreAddAttendeeExistingApptProperty
                         T_ExistingAppt
                              C_ApptByAttribute
                                   ApptByAttributeHead meeting
                                   ApptByAttributeProperties
                                        ApptByAttributeProperty
                                             ApptByAttributePeopleProperty
                                                  PreApptByAttributePeopleProperty
                                                  T_People Alex
```

is bracketed by an optional preamble and postamble (rule 5). The heads, slot preambles and postambles are originally placeholders (.*). Some placeholders are specified with part-of-speech constraints—e.g., head must be a NN (noun). For a *command* like AddAttendee, the template starts with a command part <AddAttendeeCmd>, followed by <AddAttendeeProperties>. The rest is very similar to that of the template rules for an entity. The template sets up the structural skeleton of a grammar. Hence the task of grammar learning becomes to learn the expressions for the pre-terminals like heads, commands, preambles, etc. The placeholders, without any learning, can match anything and result in ambiguities. When the learned grammar is used in our experiments, the placeholders were allowed to match any input string with a large penalty. The non-terminal <T_People> is application dependent and therefore will be provided by the developer (in the form of a name list in this example).

where the pre-terminals in *italic* are place-holders that are not matched with any word in the sentence, because neither the template grammar nor the annotation provides sufficient information for the correct decision. The terminals in bold face are matched to the pre-terminals according to the template grammar or the annotation. For example, *Ed* and *Alex* are respectively attached to the two T_People positions due to the constraints from the annotation. *Meeting* is associated with ApptByAttributeHead because it is the only remaining NN found by POS tagger and the template grammar requires that the head be a NN. *Invite*, which appears in front of *Ed*, can match both AddAttendeeCmd and PreAddAttendeePeopleProperty. Since the latter is optional, it is matched against the former. The remaining words, *to*, *the*, and *with*, cannot be deterministically aligned to any pre-terminals by the parser. However, given the partial parse tree, they can only be aligned with those pre-terminals in italic.

This effectively reduces the search space for possible alignment.

5.4.2.3. Specializing the Inherited Template CFG by Alignment of Pre-Terminal and Text.   After obtaining the parse tree with the template grammar and the annotation for the above example sentence, syntactic clues are then used to align the remaining words that are not covered in the parse tree. Prepositions and determiners can only combine with the word behind them, hence "to the" cannot align with the pre-terminal *PostAddAttendeePeopleProperty*. This leaves *PreAddAttendeeExistingApptProperty* the only choice.For the same reason, *PreApptByAttributeStartTimeProperty* is the only pre-terminal that *with* has to be aligned with. Therefore we can induce the following rules (which will be added to the existing template CFG rules):

PreAddAttendeeExistingApptProperty → *to the*
PreApptByAttributePeopleProperty. → *with*

Sometimes syntactic clues are not enough to resolve all the ambiguities. In this case, the system prompts the developer for the right decision. We are currently working on an alignment model based on the Expectation-Maximization algorithm to do this automatically.

6.   MiPad User Interface Design and Evaluation

MiPad takes advantage of the graphical display in the UI design. The graphical display simplifies dramatically the dialog management. For instance, MiPad is able to considerably streamline the confirmation and error repair strategy as all the inferred user intentions are confirmed *implicitly* on the screen. Whenever an error occurs, the user can correct it in different modalities, either by soft keyboard or speech. The user is not obligated to correct errors immediately after they occur. The display also allows MiPad to confirm and ask the user many questions in a single turn. Perhaps the most interesting usage of the display, however, is the *Tap & Talk* interface.

6.1.   Tap & Talk Interface

Because of MiPad's small form-factor, the present pen-based methods for getting text into a PDA (Graffiti, Jot, soft keyboard) are potential barriers to broad market acceptance. Speech is generally not as precise as mouse or pen to perform position-related operations. Speech

Table 2.   Complementary strengths of pen and speech as input modalities.

Pen	Speech
Direct manipulation	Hands/eyes free manipulation
Simple actions	Complex actions
Visual feedback	No visual feedback
No reference ambiguity	Reference ambiguity

interaction can also be adversely affected by the unexpected ambient noise, despite the use of denoising algorithms in MiPad. Moreover, speech interaction could be ambiguous without appropriate context information. Despite these disadvantages, speech communication is not only natural but also provides a powerful complementary modality to enhance the pen-based interface if the strengths of using speech can be appropriately leveraged and the technology limitations be overcome. In Table 2, we elaborate several cases which show that pen and speech can be complementary and used effectively for handheld devices. The advantage of pen is typically the weakness of speech and vice versa.

Through usability studies, we also observe that users tend to use speech to enter data and pen for corrections and pointing. Three examples in Table 3 illustrate that MiPad's *Tap and Talk* interface can offer a number of benefits. MiPad has a field that is always present on the screen as illustrated in MiPad's start page in Fig. 7(a) (the bottom gray window is always on the screen).

Tap & Talk is a key feature of the MiPad's user interface design. The user can give commands by tapping the *Tap & Talk* field and talking to it. *Tap & Talk* avoids speech detection problem that are critical to the noisy environment deployment for MiPad. The appointment form shown on MiPad's display is similar to the underlying semantic objects. By tapping to the *attendees* field in the calendar card shown in Fig. 7(b), for example, the semantic information related to potential attendees is used to constrain both CSR and SLU, leading to a significantly reduced error rate and dramatically improved throughput. This is because the perplexity is much smaller for each slot-dependent language and semantic model. In addition, *Tap & Talk* functions as a user-initiative dialog-state specification. The dialog focus that leads to the language model is entirely determined by the field tapped by the user. As a result, even though a user can navigate freely using the stylus in a pure GUI mode, there is no need for MiPad to include any special mechanism to handle spoken dialog focus and digression.

Table 3.   Three examples showing benefits to combine speech and pen for MiPad user interface.

Actions	Benefits
Ed uses MiPad to read an e-mail, which reminds him to schedule a meeting. Ed taps to activate microphone and says *Meet with Peter on Friday.*	Using speech, information can be accessed directly, even if not visible. Tap and talk also provides increased reliability for ASR.
Ed taps Time field and says *Noon to one thirty*	Field values can be easily changed using field-specific language models
Ed taps Subject field dictates and corrects the text about the purpose of the meeting.	Bulk text can be entered easily and faster.

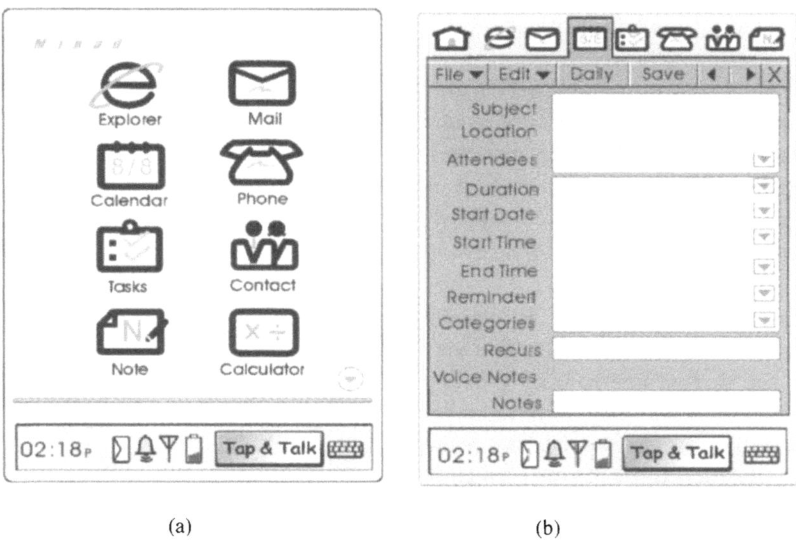

Figure 7.   Concept design for (a) MiPad's first card and (b) MiPad's calendar card.

6.2.   *Back Channel Communications*

MiPad handles back-channel communications on the device. As a user speaks, it displays a graphical meter reflecting the volume of the recording. When the utterance is beyond the normal dynamic range, red bars are shown to instruct the user to tone down. As the host computer processes the user's utterance, a running status bar is shown. The user can click a cancel button next to the status bar to stop the processing at the host computer. If the status bar vanishes without changing the display, it indicates the utterance has been rejected either by the recognizer or by the understanding system. MiPad's error repair strategy is entirely user initiative: the user can decide to try again or do something else.

6.3.   *User Study Results*

Our ultimate goal is to make MiPad produce real value to users. It is necessary to have a rigorous evaluation to measure the usability of the prototype. Our major

concerns are:

- *"Is the task completion time much better?"* and
- *"Is it easier to get the job done?"*

For our user studies, we set out to assess the performance of the current version of MiPad (with PIM features only) in terms of task-completion time, text throughput, and user satisfaction. In this evaluation, computer-savvy participants who had little experience with PDAs or speech recognition software used the partially implemented MiPad prototype. The tasks we evaluated include creating a new appointment and creating a new email. Each participant completed half the tasks using the tap and talk interface and half the tasks using the regular pen-only iPaq interface. The ordering of tap and talk and pen-only tasks is statistically balanced.

6.3.1. *Is the Task Completion Time Much Better?*
Twenty subjects were included in the experiment to evaluate the tasks of creating a new email, and creating

a new appointment. Task order was randomized. We alternated tasks for different user groups using either pen-only or *Tap & Talk* interfaces. The text throughput is calculated during e-mail paragraph transcription tasks. On average it took the participants 50 seconds to create a new appointment with the *Tap & Talk* interface and 70 seconds with the pen-only interface. This result is statistically significant with $t(15) = 3.29$, $p < .001$. The saving of time is about 30%. For transcribing an email it took 2 minutes and 10 seconds with *Tap & Talk* and 4 minutes and 21 seconds with pen-only. This difference is also statistically significant, $t(15) = 8.17$, $p < .001$. The saving of time is about 50%. Error correction for the *Tap & Talk* interface remains as one of the most unsatisfactory features. In our user studies, calendar access time using the *Tap & Talk* methods is about the same as pen-only methods, which suggests that pen-based interaction is suitable for simple tasks.

6.3.2. Is It Easier to Get the Job Done?   Fifteen out of the 16 participants in the evaluation stated that they preferred using the *Tap & Talk* interface for creating new appointments and all 16 said they preferred it for writing longer emails. The preference data is consistent with the task completion times. Error correction for the *Tap & Talk* interface remains as one of the most unsatisfactory features. On a seven point Likert scale, with 1 being "disagree" and 7 being "agree", participants responded with a 4.75 that it was easy to recover from mistakes.

Figure 8 summarizes the quantitative user study results on task completion times of email transcription and of making appointment, showing comparisons of the pen-only interface with the *Tap & Talk* interface. The standard deviation is shown above the bar of each performed task.

7.   Summary and Future Work

Speech is a necessary modality to enable a pervasive and consistent user interaction with computers across different devices—large or small, fixed or mobile, and it has the potential to provide a natural user interaction model. However, the ambiguity of spoken language, the memory burden of using speech as output modality on the user, and the limitations of current speech technology have prevented speech from becoming the choice of mainstream interface. Multimodality is capable of dramatically enhancing the usability of speech interface because GUI and speech have complementary

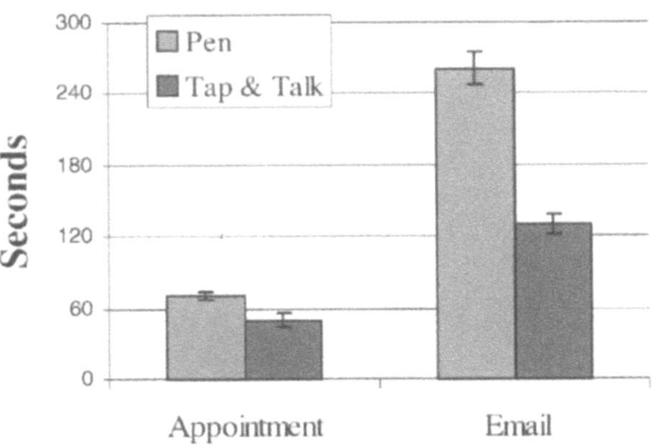

Figure 8.   User Study on task completion times of email transcription and of making appointment, showing comparisons of the pen-only interface with the Tap and Talk interface. The standard deviation is shown above the bar of each performed task.

strengths as we have shown in this paper. Multimodal access will enable users to interact with an application in a variety of ways—including input with speech, keyboard, mouse and/or pen, and output with graphical display, plain text, motion video, audio, and/or synthesized speech. Each of these modalities can be used independently or simultaneously.

Dr. Who is Microsoft's attempt to develop a speech-centric multimodal user interface framework and its enabling technologies. MiPad as we have focused on in this paper is the first *Dr. Who's* application that addresses specifically the mobile interaction scenario and it aims at the development of a consistent human-computer interaction model and component technologies for multimodal applications. Our current applications comprise mainly the PIM functions. Despite its current incomplete implementation, we have observed that speech and pen have the potential to significantly improve user experience in our preliminary user study. Thanks to the multimodal interaction, MiPad also offers a far more compelling user experience than standard voice-only telephony interaction.

The success of MiPad depends on spoken language technology and an always-on wireless connection. With upcoming 3G wireless deployments in sight, the critical challenge for MiPad remains the accuracy and efficiency of our spoken language systems since likely MiPad will be used in the noisy environment with no availability of a close-talking microphone, and the server also needs to support a large number of MiPad clients.

To meet this challenge, much of our recent work has focused on: (1) noise-robustness and transmission efficiency aspects of the MiPad system in the distributed speech processing environment, and (2) SLU with specific attention paid also to robustness as well as to automatic and high-quality application grammar development. We report our new front-end speech processing algorithm developments and some related evaluation results in this paper. Various other MiPad system components are also presented, including HMM-based speech modeling, a unified language model that integrates CFGs and N-grams for speech decoding, several key aspects of SLU (schema-based knowledge representation for the MiPad's PIM functionality, a unified statistical framework for ASR/SLU/dialogue, the robust chart parser, and semi-automatic MiPad grammar learning), *Tap & Talk* multimodal user interface, error repair strategy, and user study results comparing the multimodal interaction and the pen-only PDA interface.

The prototype of MiPad as the first *Dr. Who* application discussed in this paper has recently been successfully transferred from the research lab to the Microsoft .NET speech product division as a client browser component in the grand Kokanee architecture. This new architecture is aimed at speech-enabling the web applications based on the Speech-Application-Language-Tag standard for multimodal (speech and GUI) interactions between end users and either mobile or fixed devices. Future research work will be focused on the next version of *Dr. Who* and its new applications aimed to provide a greater degree of intelligence and automaton to larger domains and tasks than the limited PIM task of the MiPad developed so far.

Note

1. Here we assume conditional independence between the speech acoustics and the semantic object given the word sequence.

References

1. A. Acero and R. Stern, "Robust Speech Recognition by Normalization of the Acoustic Space," in *Proc. ICASSP*, Toronto, 1991.
2. X.D. Huang et al., "MiPad: A Next Generation PDA Prototype," in *Proc. ICSLP-2000*, Beijing, China, Oct. 2000.
3. K. Wang, "Implementation of a Multimodal Dialog System Using Extended Markup Language," in *Proc. ICSLP-2000*, Beijing, China, 2000.
4. L. Deng, A. Acero, L. Jiang, J. Droppo, and X.D. Huang. "High-Performance Robust Speech Recognition Using Stereo Training Data," in *Proc. ICASSP-2000*, Salt Lake City, Utah, 2001, vol. I, pp. 301–304.
5. X.D. Huang et al. "MiPad: A Multimodal Interaction Prototype," in *Proc. ICASSP-2001*, Salt Lake City, Utah, 2001, vol. I, pp. 9–12.
6. Y. Wang, M. Mahajan, and X. Huang, "A Unified Context-Free Grammar and N-Gram Model for Spoken Language Processing," in *Proc. ICASSP-2000*, Istanbul, Turkey, 2000.
7. Y. Wang, "A Robust Parser for Spoken Language Understanding," in *Proc. Eurospeech-1999*, Budapest, Hungary, 1999.
8. J. Droppo, L. Deng, and A. Acero, "Evaluation of the SPLICE Algorithm on the Aurora2 Database," in *Proc. EuroSpeech-2001 (web update version)*, Aalborg, Demark, 2001.
9. L. Deng, J. Droppo, and A. Acero. "Recursive Estimation of Nonstationary Noise Using a Nonlinear Model with Iterative Stochastic Approximation," in *Proc. IEEE Workshop on ASRU-2001*, Italy, Dec. 2001.
10. Y. Wang and A. Acero, "Grammar Learning for Spoken Language Understanding," in *Proc. IEEE Workshop on ASRU-2001*, Madonna di Campiglio, Italy, Dec. 2001.
11. H.G. Hirsch and D. Pearce, "The AURORA Experimental Framework for the Performance Evaluations of Speech Recognition Systems Under Noisy Conditions," *ISCA ITRW ASR2000 "Automatic Speech Recognition: Challenges for the Next Millennium,'* Paris, France, Sept. 18–20, 2000.
12. L. Deng, J. Droppo, and A. Acero. "Robust Speech Recognition Using Iterative Stochastic Approximation and Recursive EM for Estimation of Nonstationary Noise," submitted to *IEEE Trans. Speech and Audio Proc.*, 2001.
13. K. Wang, H. Hon, and A. Acero, "A Distributed Understanding and Dialog Environment Using Web Infrastructure," submitted to *IEEE Trans. Speech and Audio Proc.*, 2001.
14. L. Deng, A. Acero, M. Plumpe, and X.D. Huang, "Large-Vocabulary Speech Recognition Under Adverse Acoustic Environments," in *Proc-ICSLP2000*, Beijing, China, Vol. 3, 2000, pp. 806–809.

Li Deng received a B.S. from the University of Science and Technology of China (Biophysics) in 1982, a Master's degree (Electrical Engineering) from the University of Wisconsin-Madison in 1984, and a Ph.D. degree (Electrical Engineering) from the University of Wisconsin-Madison in 1986.

He worked on large vocabulary automatic speech recognition in Montreal, Canada, 1986–1989. In 1989, he joined the Dept. of Electrical and Computer Engineering at the University of Waterloo, Ontario, Canada as Assistant Professor, where he became tenured Full Professor in 1996. From 1992 to 1993, he conducted sabbatical research at Laboratory for Computer Science, Massachusetts Institute of Technology, Cambridge, Mass, and from 1997–1998, at

ATR Interpreting Telecommunications Research Laboratories, Kyoto, Japan. In Dec. 1999, he joined Microsoft Research, Redmond, WA as Senior Researcher. His research interests include acoustic-phonetic modeling of speech, speech and speaker recognition, speech synthesis and enhancement, speech production and perception, auditory speech processing, noise robust speech processing, statistical methods and machine learning, nonlinear signal processing, spoken language systems, multimedia signal processing, and multimodal human-computer interaction. In these areas, he has published over 200 technical papers and book chapters, and has given keynote, tutorial, and other invited lectures. He recently completed the book "Speech Processing—A Dynamic and Optimization-Oriented Approach" (2003, Marcel Dekker Publishers, New York).

Dr. Deng served on Education Committee and Speech Processing Technical Committee of the IEEE Signal Processing Society during 1996–2000, and is currently serving as Associate Editor for IEEE Transactions on Speech and Audio Processing. He is Technical Chair of the 2004 International Conference on Acoustics, Speech and Signal Processing (ICASSP).
deng@microsoft.com

Ye-Yi Wang (M'99) received the B.Eng. and M.S. degree in Computer Science and Engineering from Shanghai Jiao Tong University in 1985 and 1987, respectively. He received the M.S. degree in Computational Linguistics and the Ph.D. degree in Language and Information Technology from Carnegie Mellon University in 1992 and 1998, respectively.

He is currently a researcher of the Speech Technology Group at Microsoft Research. His research interests include spoken language understanding, language modeling, machine translation and machine learning.

Kuansan Wang received BS from National Taiwan University in 1986, MS and PhD from University of Maryland at College Park in 1989 and 1994, respectively, all in Electrical Engineering. From 1994 to 1996, he was with the Speech Research Department at Bell Labs, Murray Hill, NJ. From 1996 to 1998, he was with speech and spoken language labs at NYNEX Science and Technology Center in

White Plains, NY. Since 1988, he has been with speech technology group at Microsoft Research in Redmond WA.

Dr. Wang's research areas are speech recognition, spoken language understanding and multimodal dialog systems.

Alex Acero received an engineering degree from the Polytechnic University of Madrid (Spain) in 1985, a master's degree from Rice University, (Houston, TX) in 1987 and a PhD from Carnegie Mellon University, (Pittsburgh, PA) in 1990, all in electrical engineering.

He was a senior voice engineer at Apple Computer (1990–1991) and manager of the speech technology group at Telefonica Investigacion y Desarrollo (1991–1993). He joined Microsoft Research, (Redmond, WA), in 1994, where he is currently manager of the speech group. He is also affiliate professor at the University of Washington (Seattle, WA). He is author of the books *Spoken Language Processing* (Upper Saddle River, NJ, Prentice Hall, 2000) and *Acoustical and Environmental Robustness in Automatic Speech Recognition* (Boston, MA, Kluwer, 1993). He also has written chapters in three edited books, 7 patents and over 50 other publications. His research interests include noise robustness, speech synthesis, signal processing, acoustic modeling, statistical language modeling, spoken language processing, speech-centric multimodal interfaces and machine learning.

Dr. Acero served in the *IEEE Signal Processing Society's Speech Technical Committee* as member (1996–2000) and chair (2000–2002). He was general co-chair of the *2001 IEEE Workshop on Automatic Speech Recognition and Understanding*, sponsorship chair of the *1999 IEEE Workshop on Automatic Speech Recognition and Understanding* and publications chair of *ICASSP 98*. He is associate editor of *Computer, Speech and Language*.

Hsiao-Wuen Hon received the B.S. in Electrical Engineering from National Taiwan University; and M.S. & PhD degrees in Computer Science from Carnegie Mellon University.

Dr. Hsiao-Wuen Hon is an Architect in Speech.Net at Microsoft Corporation. Prior to his current position, Dr. Hon had been a senior researcher at Microsoft Research and has been a key contributor of Microsoft's Whisper and Whistler technologies, which are the corner stone for Microsoft SAPI and SDK product family. Before joining Microsoft, Dr. Hon worked at Apple Computer, Inc., where he was a principal researcher and technology supervisor at Apple-ISS research center. Dr. Hon led the research and development for Apple's' Chinese Dictation Kit, which received excellent reviews from many industrial publications and a handful of rewards, including Comdex Asia'96 Best software product medal, Comdex Asia'96 Best of the Best medal and Singapore National Technology award. While at CMU, Dr. Hon is the co-inventor of CMU SPHINX system on which many commercial speech recognition systems are based on, including Microsoft and Apple. Hsiao-Wuen is an international recognized speech technologist and has published more than 70 technical papers in various international journals and conferences.

He is currently a senior member of IEEE and an associated editor for IEEE Transaction of Speech and Audio Processing. He co-authored with X.D. Huang and Alex Acero on a book titled *Spoken Language Processing*. He has also been serving as reviewers and chairs for many international conferences and journals. Dr. Hon holds 9 US patents and currently has 9 pending patent applications.

James "Jasha" Garnet Droppo III received the Bachelor of Science in Electrical Engineering degree, *cum laude*, honors, from Gonzaga University in 1994. He received the M.S. degree in Electrical Engineering and Ph.D in Electrical Engineering from the University of Washington under Les Atlas in 1996 and 2000, respectively. At the University of Washington, he helped to develop and promote a discrete theory for time-frequency representations of audio signals, with a focus on speech recognition. He joined the Speech Technology Group at Microsoft Research in 2000. His academic interests include noise robustness and feature normalization for speech recognition, compression, and time-frequency signal representations.

Constantinos Boulis is currently a PhD student in the University of Washington, Seattle. Prior to that he obtained a M.Sc. degree from the Computer Engineering Departmentof Technical University of Crete, Greece from where he also holds an undergraduate degree. His academic interests include unsupervised topic detection in unconstrained speech, distributed speech recognition, speaker adaptation and pattern recognition in general.

Milind Mahajan received B.Tech. degree in Computer Science and Engineering from Indian Institute of Technology in Mumbai, India in 1986. He received M.S. degree in Computer Science from University of Southern California in 1988. He is currently a researcher in the Speech Technology Group of Microsoft Research. His research interests include language modeling and machine learning.

X.D. Huang as general manager of Microsoft. NET Speech, Dr. Xuedong (XD) Huang is responsible for the development of Microsoft's speech technologies, speech platform, and speech development tools. .NET Speech includes .NET Speech Application Server, .NET Speech Components, .NET Speech SDK, and Speech Technology Research groups. Huang is widely known for his pioneering work in the areas of spoken language processing. Huang and his team have created core technologies used in a number of Microsoft's products including both Office XP and Windows XP, and pioneered the industry-wide SALT initiatives.

Xuedong (XD) Huang joined Microsoft Research as a Senior Researcher to lead the formation of Microsoft's Speech Technology Group in 1993. Prior to joining Microsoft, he was on the faculty of Carnegie Mellon's School of Computer Sciences and directed the effort in developing CMU's Sphinx-II speech recognition system.

Huang is on the board of SALT Forum. He is also an affiliate Professor of Electrical Engineering at University of Washington, and an honorary professor of Computer Science at his *alma mater* Hunan University. He has published more than 100 journal and conference papers and is a frequent keynote speaker in numerous industry conventions. He has co-authored two books: Hidden Markov Models for Speech Recognition (Edinburgh University Press 1990) and Spoken Language Processing (Prentice Hall 2001). Huang's professional awards include National Education Commission of China's 1987 Science and Technology Progress Award, IEEE Signal Processing Society's 1992 Paper Award, and Allen Newell Research Excellence Medal.

Huang holds a doctorate in Electrical Engineering from University of Edinburgh, a master's in Computer Sciences from Tsinghua University, and a bachelor's in Computer Sciences from Hunan University. Huang is a Fellow of the IEEE.

Journal of VLSI Signal Processing 36, 189–203, 2004

Blind Model Selection for Automatic Speech Recognition in Reverberant Environments*

LAURENT COUVREUR

Multitel—TCTS, Faculté Polytechnique de Mons, 1 Avenue Copernic, B-7000 Mons, Belgium

CHRISTOPHE COUVREUR

Speech & Language Technology Division, Scansoft, Inc., 32 Guldensporenpark, B-9820 Merelbeke, Belgium

Received December 26, 2001; Revised July 8, 2002; Accepted July 18, 2002

Abstract. This communication presents a new method for automatic speech recognition in reverberant environments. Our approach consists in the selection of the best acoustic model out of a library of models trained on artificially reverberated speech databases corresponding to various reverberant conditions. Given a speech utterance recorded within a reverberant room, a Maximum Likelihood estimate of the fullband room reverberation time is computed using a statistical model for short-term log-energy sequences of anechoic speech. The estimated reverberation time is then used to select the best acoustic model, i.e., the model trained on a speech database most closely matching the estimated reverberation time, which serves to recognize the reverberated speech utterance. The proposed model selection approach is shown to improve significantly recognition accuracy for a connected digit task in both simulated and real reverberant environments, outperforming standard channel normalization techniques.

Keywords: room reverberation, maximum likelihood estimation, automatic speech recognition

1. Introduction

During the past decade, automatic speech recognition (ASR) has been successfully deployed in desktop applications (e.g., dictation systems) and in interactive telephone-based systems (e.g., voice portals). These systems typically operate with a close-talking microphone in quiet environments. Nowadays, we observe a growing interest for voice-controlled devices which can be used in more natural conditions, the so-called *hands-free* devices [1]. However, there still exist important obstacles to the mass development of such devices. The poor accuracy of the current ASR systems in realistic hands-free operating conditions is the main one.

Today's state-of-art ASR systems are based on stochastic models of speech acoustics. That is, they extract a set of significant acoustic features from the speech signal and use statistical models to represent the distribution of these features for speech units such as words, syllables or phonemes. Recognition of an unknown utterance is then treated as a statistical pattern recognition problem. These statistical models, often referred to as acoustic models, have to be trained on large speech databases. Unfortunately, the performance of the ASR systems degrade dramatically when they are presented with speech having different acoustic features than the ones statistically modeled during the training procedure. For hands-free devices, the speech signal propagates from the speaker to a distant microphone in an unknown acoustic environment. In realistic acoustic environments, the speech signal is contaminated by competing noise sources and distorted by room reverberation before reaching the microphone.

*This work was supported in part by F.N.R.S. (Fonds National de la Recherche Scientifique) with a F.R.I.A. grant (Fonds pour la formation à la Recherche dans l'Industrie et l'Agriculture, Belgium).

Since the ASR acoustic models are usually trained on noise-free anechoic speech, they are unlikely to match the distribution of acoustic features of the recorded speech in a noisy reverberant environment.

In this communication, we are primarily concerned by the effect of room reverberation. When an acoustic signal (a speech signal) is generated in a reverberant room, it follows multiple paths from the source (the speaker) to the receiver (the microphone). While a version of the source signal propagates directly to the receiver through the air, other delayed versions arrive after bouncing back and forth on the reflective surfaces of the room. The phenomenon severely affects the spectral characteristics of the acoustic signal picked up at the microphone. It globally results in a "smearing" effect of the acoustic energy forward in time. Figure 1 shows the waveforms and the corresponding spectrograms of an anechoic speech utterance recorded with a close-talking microphone and its reverberated version recorded with a distant-talking microphone in a typical reverberant room.

The distortion caused by room reverberation is especially harmful for ASR systems [2–4]. Several approaches have been proposed to reduce the discrepancy between the training conditions (close-talking/anechoic speech) and the operating conditions (distant-talking/reverberated speech) due to room reverberation. These methods can be roughly separated into three categories:

Speech Enhancement. A first approach consists in attempting to recover an "enhanced" speech signal close to the anechoic one by processing its reverberated version while keeping the acoustic models unchanged. These methods may be further categorized into single-microphone methods [5–9] and multi-microphone methods [10–14]. While these methods can produce high quality enhanced speech, they are generally computationally demanding and may require solving ill-conditioned mathematical problems [15, 16].

Model Adaptation. Other methods propose to reduce the mismatch between the training and the operating conditions by adapting the acoustic models from observed reverberated speech. Many efficient adaptation schemes [17–22] have been proposed to address the problem of ASR in reverberant environments. However, they generally require a significant amount of adaptation data, thereof limiting their practical use in realistic conditions where the acoustic environment is rapidly changing.

Robust Features. The last class of methods includes all the preprocessing parts of the ASR systems, the so-called front-ends, that extract acoustic features of speech insensitive to channel distortion [23–29]. These methods have the advantage of being easy to implement. They are cost-effective since there is no need to adapt the acoustic models given reverberated observations or to restore the speech signal prior to

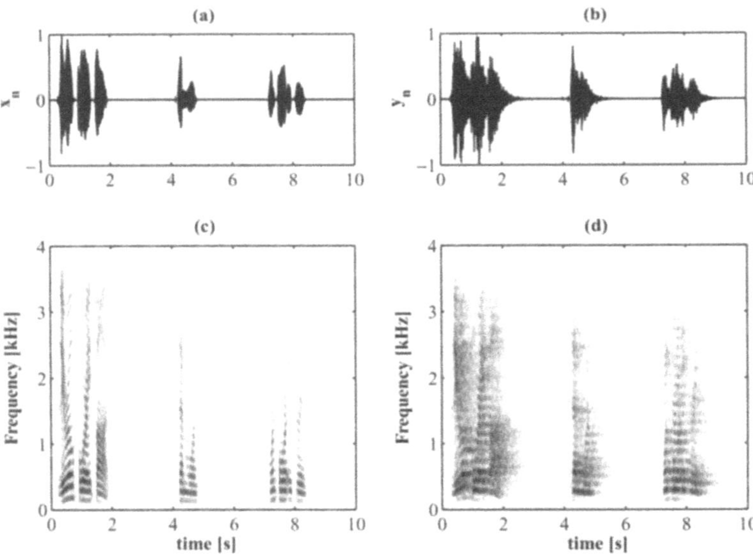

Figure 1.   Waveforms of (a) an anechoic speech utterance x_n (digit sequence "z485-71-1483") and (b) its reverberated version y_n; (c) and (d) show the respective spectrograms.

computing the acoustic features. However, they often perform poorly in moderately to highly reverberant environments.

In this communication, we investigate a new technique that can be somewhat regarded as a model adaptation method. In our approach, a library of acoustic models are trained separately on speech databases corresponding to various reverberant conditions. During operation, the best model is selected and used to recognize the incoming reverberated speech. The best model is actually the model that most closely matches the operating reverberant conditions. Unlike in classical adaptation techniques, the speech data to be recognized are not used to adapt the acoustic model but simply to select one from the library. This method is fast since it requires little data for "adapting" the acoustic model by selection from the library. It is also computationally efficient since the model "adaptation" is actually done beforehand during the training procedure. In order to implement this approach, two issues have to be addressed: the procedure for creating the library of acoustic models and the procedure for selecting a model from the library during operation,

Training Procedure. Ideally, a training database should be collected for every possible operating reverberant environment where the ASR system is to be deployed. Hence, numerous large speech databases would need to be collected in order to cover most of reverberant conditions. The problem is further compounded if the ASR system must be speaker-independent and if several languages must be supported. The database collection effort rapidly becomes unmanageable.

Alternatively, the reverberated speech databases may be obtained by adequately reverberating an anechoic speech database. It is commonly assumed that the transmission of an acoustic signal from a source to a microphone within a reverberant enclosure may be modeled by a linear filter [33]. Under this assumption, a reverberated speech database may be obtained by convolving an anechoic speech database with a room impulse response measured between the source and the microphone in the operating reverberant environment [17, 30, 31]. However, this approach ignores that room impulse responses are highly dependent on the acoustic characteristics of the reverberant enclosure (e.g., a room impulse response changes when the temperature and the humidity vary

or when air currents shift [32]), on the geometric characteristics of the room (e.g., a room impulse response changes when doors are opened or closed), on the source and the microphone locations (e.g., a room impulse response changes when the speaker moves around [15]), etc. It is thus extremely hard to guarantee that the measured room impulse response would exactly match the room impulse response of the operating reverberant environment. There is a risk that the ASR system will then "over-fit" the reverberant conditions observed while measuring the room impulse response, leading to disappointing results when the operating conditions change, even slightly [3].

To resolve this problem, we proposed in [3] to use a "randomized" synthetic reverberating filter instead of a measured room impulse response to obtain reverberated speech databases. The impulse response for the synthetic reverberating filter is generated randomly multiple times during the reverberation process of the anechoic speech database. It is constrained to match a high-level, perceptually meaningful, acoustic property of the operating reverberant environment, namely the fullband reverberation time T_{60} [33]. The randomization of the low-level details of the impulse response is expected to model the effects of possible speaker/microphone movements and of small variations in enclosure geometry or acoustic characteristics, consequently avoiding over-fitting the acoustic models to a particular room impulse response. Artificially reverberated speech databases (and the acoustic models thereof) can be generated for various reverberation times to build the desired library of acoustic models. This approach requires neither collecting large speech databases nor measuring room impulse responses. While the assumption that the fullband reverberation time T_{60} is sufficient to capture the properties of a reverberant environment may seem overly simplistic, it has been found to yield good results in the framework of ASR. The complete training procedure is presented in details in [3].

Selection Procedure. Given a reverberated speech utterance to be recognized, we want to select the best acoustic model out of the library generated with the training procedure described above. In our approach, it is sufficient to estimate the fullband reverberation time T_{60} for the speech utterance and then select an acoustic model accordingly. This communication describes an algorithm for the blind estimation of T_{60} given a speech utterance recorded with a distant-talking microphone in a reverberant environment. The proposed algorithm is inspired by the Stochastic Matching concept introduced by Sankar and Lee in [34]. Its basic principles can be briefly summarized as follows.

The impact of room reverberation is defined in the short-time log-energy domain by a parametric model whose parameters are related to the reverberation time. Although this model relies on a strong assumption, namely that room reverberation can be entirely characterized by the fullband reverberation time T_{60}, it allows us to obtain a mathematically tractable model and, as will be seen later, it works satisfyingly in practice. Given a reverberated speech utterance and its corresponding short-term log-energy sequence, the estimation algorithm computes a Maximum Likelihood (ML) estimate of the parameters of the distortion model, and an estimate of the reverberation time is derived. This algorithm requires a statistical model of short-term log-energy sequences for anechoic speech that must be trained beforehand on an anechoic speech database.

In the next section, we define the simplified room reverberation model used for the selection procedure. Then, the algorithm for the blind estimation of T_{60} is presented in Section 3. In Section 4, we report experimental results for the recognition of connected digit sequences. For these experiments, our model selection procedure is tested with a library of acoustic models built with artificially reverberated speech databases. Results are given for both simulated and real reverberant environments. In the former case, test material is obtained by simulating a rectangular room with the Image Method [35, 36]. In the latter case, test recordings have been collected in various real reverberant enclosures. The model selection approach is compared with standard channel compensation techniques [23, 24]. Although the hands-free ASR problem is addressed only partially in this communication (i.e., reverberant but noise-free conditions are considered), the method shows promising results. The results obtained in Section 4 allow us to identify limitations of the current approach. Conclusions and directions for future research to address these shortcomings are given in Section 5.

2. Room Reverberation Model

The propagation between a source and a microphone within a reverberant enclosure is classically modeled in the time domain by a linear filter. A reverberated speech waveform y_n is obtained by convolving its anechoic version x_n with a causal room impulse response h_n,

$$y_n = x_n * h_n. \tag{1}$$

As mentioned above, room impulse responses are highly dependent on the acoustic and geometric characteristics of the room, and on the source and microphone locations. However, under the diffuse sound field assumption [33], it can be shown that the ensemble average ε_n of the squared room impulse responses is a decaying exponential:

$$\varepsilon_n \stackrel{\triangle}{=} \langle h_n^2 \rangle = \varepsilon_0 e^{-kn}, \tag{2}$$

where the damping constant k is related to the fullband reverberation time T_{60} (s) of the reverberant room by

$$k = \log 10^6 / (T_{60} \times F_s) \tag{3}$$

with F_s (Hz) standing for the sampling frequency. Equation (3) is based on the classical definition of the reverberation time [33]: T_{60} is the time interval in which the sound energy within a reverberant room reaches one millionth (i.e., -60 dB) of its initial value once the sound source has been interrupted. These equations suggest that it is possible to generate a synthetic room impulse response matching a given reverberation time by modulating a zero-mean random sequence with a decaying exponential [37]. This forms the basis for the technique proposed in [3] to generate artificially reverberated speech databases by filtering an anechoic speech database in order to train acoustic models in various reverberant conditions, i.e., for various reverberation times.

Room reverberation is fully described by the room impulse response between the speaker and the microphone. In order to estimate the reverberation time T_{60} for selecting an acoustic model, one can propose to identify blindly the room impulse response from recorded reverberated speech, and then to compute the reverberation time from the estimated impulse response, for instance, using Schroeder's method [33]. But since the blind identification of a room impulse response is a difficult ill-conditioned problem, we propose to use instead a coarser model of room reverberation to simplify the T_{60} estimation problem. We decide to model the impact of room reverberation on short-term energy sequences instead of on time-domain speech signals. The short-term energy sequence W_m of an anechoic speech signal x_n (the speaker voice) is obtained by slicing it into (possibly overlapping) frames, and by computing the energy of every frame. For the m-th frame, we have

$$W_m \triangleq \frac{1}{N_w} \sum_{n=mN_r}^{mN_r+N_w-1} x_n^2, \qquad (4)$$

where $N_w \triangleq T_w \times F_s$ and $N_r \triangleq F_s/F_r$ with T_w (s) and F_r (Hz) denoting the frame length and the frame rate, respectively. Likewise, we define Z_m as the short-term energy sequence of the reverberated speech signal y_n (the microphone signal). Under the assumption that the sound field is diffuse and that Eq. (2) is valid, the energy of the m-th frame W_m will smear forward in time, its contribution to the following frames decaying exponentially. We assume that this smearing effect in the short-term energy domain can be modeled by a first order auto-regressive (AR) filter:

$$\alpha_0 Z_m = W_m - \alpha_1 Z_{m-1}, \qquad (5)$$

with $\alpha_0 > 0$ and $\alpha_1 < 0$.

Figure 2 summarizes the reverberation model for short-term log-energy sequences and its relation to the original signals. Figure 3 gives an example of an anechoic speech utterance x_n and its reverberated version y_n recorded with a close-talking microphone and a distant-talking microphone, respectively, in a typical reverberant room. The figure also shows the distortion on the corresponding short-term log-energy sequences $X_m \triangleq 10 \log_{10} W_m$ and $Y_m \triangleq 10 \log_{10} Z_m$ computed after proper normalization of the speech signals. The short-term energy sequences are represented in the logarithmic domain instead of the linear domain because of better conditioned dynamics there. In Fig. 3, we observe that the decay of Y_m from peak to valley is almost

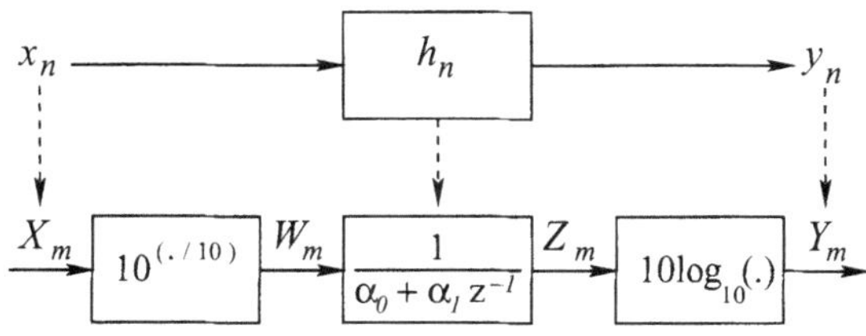

Figure 2. Room reverberation model (upper part) for temporal signals and equivalent diffuse model (lower part) for short-term log-energy sequences.

linear. The decay is thus exponential in the short-term linear-energy domain, which validates the model postulated in Eq. (5). Like in Eq. (3), the decaying rate of the short-term energy can be related to the reverberation time T_{50} by

$$T_{60} = \log 10^6/(-\log(-\alpha_1) \times F_r). \qquad (6)$$

3. Blind T_{60} Estimation Algorithm

We present here an algorithm for blindly estimating the parameters (α_0, α_1) of the reverberation model given a reverberated utterance y_n. Once α_1 has been estimated, T_{60} of the reverberant room can be derived via Eq. (6). The algorithm is blind because no other information (e.g., silence/speech segmentation or close-talking reference of the speech signal) other than a speech utterance recorded with a distant-talking microphone are needed. First, a statistical model for short-term log-energy sequences of anechoic speech utterances is introduced. Next, an estimation algorithm using this model is described. Finally, some results to demonstrate its convergence properties are reported.

3.1. Model for Speech Production

Typical short-term log-energy sequences for anechoic speech are non-stationary and characterized by two states, called the silence and speech states. Furthermore, successive values in a given state are undoubtedly not statistically independent (see Fig. 3(c)). Hence, we decide to model short-term log-energy sequences for anechoic speech by a 2-state one-dimensional Linear Predictive Hidden Markov Model (LP-HMM) [38, 39]. In this model, a short-term log-energy sequence X_m for anechoic speech is generated by processing an emission sequence E_m with an AR filter of order P,

$$\beta_0(s_m)X_m = E_m - \sum_{p=1}^{P} \beta_p(s_m)X_{m-p}, \qquad (7)$$

whose coefficients $\beta_p(s_m)$, $p = 0, \ldots, P$, vary in time according to the evolution of a hidden Markov chain. The emissions E_m are assumed to be conditionally independent given the LP-HMM state sequence s_m and to have a Gaussian distribution with mean μ_i and standard deviation σ_i for state $s_m = i, i = \{0, 1\}$. To complete our model, we define the transition probabilities $a_{ij} \triangleq P[s_m = j|s_{m-1} = i]$ with $i, j = \{0, 1\}$.

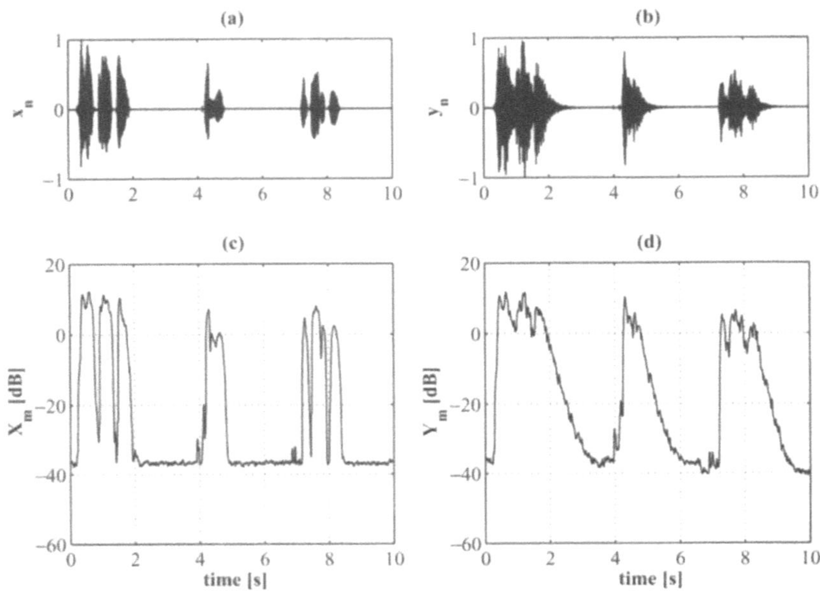

Figure 3. Waveforms of (a) an anechoic speech utterance x_n (digit sequence "z485-71-1483") and (b) its reverberated version y_n; (c) and (d) show the respective short-term log-energy sequences X_m and Y_m for $T_w = 30$ ms and $F_r = 100$ Hz.

Figure 4 illustrates a 2-state one-dimensional LP-HMM with AR filters limited to first order (i.e., $P = 1$), which is used in this work for modeling short-term log-energy sequences of anechoic speech utterances. Note that the LP-HMM is defined for log-energy sequences instead of linear-energy sequences because these data have better conditioned dynamics, and their statistical distributions are easier to model. All the parameters of the LP-HMM can be estimated by an Expectation-Maximization (EM) algorithm [39, 40]. Table 1 gives the parameters of the LP-HMM used in this work. The model has been trained on a noise-free part of the AURORA speech database [41]. Note that each speech

utterance is normalized with respect to its energy prior to computing its short-term log-energy sequence.

3.2. Algorithm Description

The proposed algorithm for estimating T_{60} is inspired by the Maximum Likelihood (ML) Stochastic Matching paradigm introduced by Sankar and Lee [34]. The Stochastic Matching method assumes that some features X are observed after a parametric distortion $F(\cdot \mid \alpha)$ such that only the distorted features $Y = F(X \mid \alpha)$ are available. It provides a framework for deriving a ML estimate of the distortion parameters α given some observed features Y and a statistical model for the unobserved features X.

Here, we consider the short-term log-energy sequences of reverberated speech utterances (microphone signal) as the observed features. The short-term

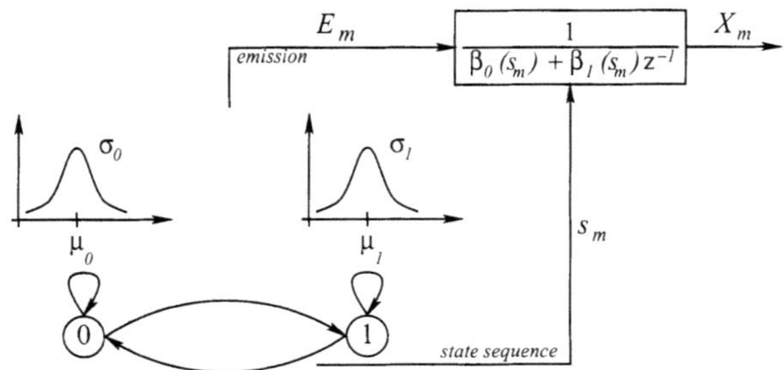

Figure 4. View of a 2-state one-dimensional first-order LP-HMM for modeling short-term log-energy sequences of anechoic speech (0: silence state, 1: speech state).

Table 1. Parameters of a 2-state one-dimensional first-order LP-HMM for modeling short-term log-energy sequences of anechoic speech ($T_w = 30$ ms and $F_r = 100$ Hz).

$s_m = i$	a_{ii}	a_{ij}	μ_i	σ_i	$\beta_0\,(i)$	$\beta_1\,(i)$
0	0.95	0.05	-4.3	4.2	1.0	-0.92
1	0.03	0.97	1.1	3.2	1.0	-0.77

log-energy sequences of the corresponding anechoic speech utterances (speaker voice) are the unobserved features. The distortion model is the room reverberation model in the short-term log-energy domain of Fig. 2. Hence, the distortion parameters are the coefficients (α_0, α_1) of the first-order AR filter of Eq. (5). The statistical model of the unobserved features is the 2-state LP-HMM described in Section 3.1.

Let $Y_{0,M} = \{Y_0, \ldots, Y_M\}$ denote the finite-length short-term log-energy sequence of a reverberated speech utterance and define $s_{0,M} = \{s_0, \ldots, s_M\}$ as the corresponding hidden state sequence of the underlying LP-HMM. We want to compute an estimate of the distortion parameters $\alpha = (\alpha_0, \alpha_1)$ based on the Stochastic Matching paradigm. That is, we want to solve the ML estimation problem

$$\hat{\alpha} = \arg\max_{\alpha} \log p(Y_{0,M} \mid \alpha) \qquad (8)$$

where $p(Y_{0,M} \mid \alpha)$ denotes the likelihood of the data $Y_{0,M}$ given the distortion parameters α. Since the underlying state sequence is unknown, we have to solve a ML problem for incomplete data. We naturally resort to an Expectation-Maximization (EM) algorithm [40] to derive a solution. The EM algorithm consists in a two-step iterative procedure.

E-step—Given current estimates $\alpha^{(\ell)}$ of the distortion parameters at iteration ℓ and some observations $Y_{0,M}$, compute the conditional expectation, also known as the auxiliary function:

$$Q(\alpha \mid \alpha^{(\ell)}) \triangleq E\big[\log p(Y_{0,M}, s_{0,M} \mid \alpha) \mid Y_{0,M}, \alpha^{(\ell)}\big]. \qquad (9)$$

M-step—Find new estimates $\alpha^{(\ell+1)}$ of the distortion parameters by maximizing Eq. (9) with respect to α:

$$\alpha^{(\ell+1)} = \arg\max_{\alpha} Q(\alpha \mid \alpha^{(\ell)}). \qquad (10)$$

Although numerical optimization techniques can be use to solve Eq. (10), we prefer deriving an explicit solution, the so-called re-estimation formulae. We first compute the first derivative of the auxiliary function with respect to α. Next, we solve for its zeros to obtain the new estimates $\alpha^{(\ell+1)}$. They are then substituted in Eq. (9) to iterate the procedure, unless convergence is reached.

In order to derive close-form re-estimation formulae, we first apply the Bayes rule and note that the state sequence does not depend on the distortion parameters. Hence, we can equivalently solve

$$\alpha^{(\ell+1)} = \arg\max_{\alpha} E\big[\log p(Y_{0,M} \mid s_{0,M}, \alpha) \mid Y_{0,M}, \alpha^{(\ell)}\big]. \qquad (11)$$

In Appendix A, we derive an approximation of the transformation $E_{0,M} = G(Y_{0,M} \mid s_{0,M}, \alpha)$ between a finite length sequence of observations $Y_{0,M}$ and the corresponding sequence of emissions $E_{0,M}$ given a state sequence $s_{0,M}$. The approximation allows us to express the likelihood function for the observations with respect to the likelihood function for the emissions [42] as

$$p(Y_{0,M} \mid s_{0,M}, \alpha) = p(E_{0,M} \mid s_{0,M}, \alpha) |\mathbf{J}|, \qquad (12)$$

where $|\mathbf{J}|$ denotes the determinant of the Jacobian matrix $\mathbf{J}$ of the transformation $G(\cdot)$. In Appendix A, we also show that $|\mathbf{J}|$ depends only on α_0 (see Eq. (33)). By proper normalization of the energy levels of the speech utterances during the training procedure of the LP-HMM and during the estimation of T_{60}, we can constrain $\alpha_0 = 1$. The maximization problem is then reduced to

$$\alpha_1^{(\ell+1)} = \arg\max_{\alpha_1} E\big[\log p(E_{0,M} \mid s_{0,M}, \alpha_1) \mid Y_{0,M}, \alpha_1^{(\ell)}\big]. \qquad (13)$$

By definition of the LP-HMM in Section 3.1, the emissions $E_{0,M}$ are conditionally independent, and Eq. (13) becomes,

$$\alpha_1^{(\ell+1)} = \arg\max_{\alpha_1} \sum_{m=0}^{M} \\ \times E\big[\log p(E_m \mid s_m, \alpha_1) \mid Y_{0,M}, \alpha_1^{(\ell)}\big]. \qquad (14)$$

The expectation operator can be factored out at ℓ-th iteration by introducing the *a posteriori* state probabilities $\gamma_{m,i}^{(\ell)} \triangleq P[s_m = i \mid Y_{0,M}, \alpha_1^{(\ell)}], i = \{0, 1\}$, yielding

$$\alpha_1^{(\ell+1)} = \arg\max_{\alpha_1} \sum_{m=0}^{M} \sum_{i=0}^{1} \gamma_{m,i}^{(\ell)} \log p(E_m \mid s_m = i, \alpha_1). \qquad (15)$$

Given the parameters of the LP-HMM, the *a posteriori* state probabilities can be efficiently computed for any sequence of observations with the Forward-Backward algorithm [39]. Using the assumption that the emissions have Gaussian distributions and that the distribution parameters are state-dependent, we find

$$\alpha_1^{(\ell+1)} = \arg\max_{\alpha_1} \sum_{m=0}^{M} \sum_{i=0}^{1} \gamma_{m,i}^{(\ell)} (E_m - \mu_i)^2 / \sigma_i^2, \quad (16)$$

where constant terms have been left out for ease of notation. Next, we can use the linear approximation of Eq. (25) derived in Appendix A and make α_1 appear explicitly:

$$\alpha_1^{(\ell+1)} = \arg\max_{\alpha_1} \sum_{m=0}^{M} \sum_{i=0}^{1} \gamma_{m,i}^{(\ell)}$$
$$\times \left(\sum_{p=0}^{1} \beta_p(s_m) \left[\sum_{k=0}^{1} \frac{\alpha_k}{\xi W_m^{(\ell)}} Z_{m-p-k} \right. \right.$$
$$\left. \left. + X_m^{(\ell)} - 1/\xi \right] - \mu_i \right)^2 \Bigg/ \sigma_i^2 \quad (17)$$

with $\xi = \log(10)/10$. The re-estimation formula for α_1 is finally obtained by setting the first-order derivative of the right-hand term of Eq. (17) with respect to α_1 to zero. Note that this first-order derivative is linear for α_1. We finally get

$$\alpha_1^{(\ell+1)} = \left(-p^{(\ell)} - q^{(\ell)} \right) / r^{(\ell)}, \quad (18)$$

with $p^{(\ell)}$, $q^{(\ell)}$ and $r^{(\ell)}$ defined by

$$c_{m,i}^{(\ell)} \triangleq \sum_{p=0}^{1} \beta_p(i) Z_{m-p-1} / \xi W_{m-p}^{(\ell)}, \quad (19)$$

$$r^{(\ell)} \triangleq \sum_{m=0}^{M} \sum_{i=0}^{1} \frac{\gamma_{m,i}^{(\ell)}}{\sigma_i^2} \left(c_{m,i}^{(\ell)} \right)^2, \quad (20)$$

$$q^{(\ell)} \triangleq \sum_{m=0}^{M} \sum_{i=0}^{1} \frac{\gamma_{m,i}^{(\ell)}}{\sigma_i^2} \left(\sum_{p=0}^{1} \frac{\beta_p(i) Z_{m-p}}{\xi W_{m-p}^{(\ell)}} \right) c_{m,i}^{(\ell)}, \quad (21)$$

$$p^{(\ell)} \triangleq \sum_{m=0}^{M} \sum_{i=0}^{1} \frac{\gamma_{m,i}^{(\ell)}}{\sigma_i^2} \left(\sum_{p=0}^{1} \beta_p(i) \left(X_{m-p}^{(\ell)} - \frac{1}{\xi} \right) - \mu_i \right)$$
$$\times c_{m,i}^{(\ell)}. \quad (22)$$

The resulting iterative algorithm for the ML estimation of the distortion parameter α_1, and therefrom the

Table 2. Iterative algorithm for the estimation of the fullband reverberation time T_{60} from the short-term log-energy sequence $Y_{0,M}$ of a reverberated speech utterance.

1. Initialize the estimate of the distortion parameter $\alpha_1^{(0)}$ ($\alpha_0 = 1$); Set $\ell = 0$; and compute $Z_m = 10^{Y_m/10}$, $m = 0, \ldots, M$.
2. Apply the current inverse distortion filter to obtain $W_m^{(\ell)} = \alpha_0^{(\ell)} Z_m + \alpha_1^{(\ell)} Z_{m-1}$; and compute $X_m^{(\ell)} = 10 \log_{10} W_m^{(\ell)}$, $m = 0, \ldots, M$.
3. Estimate the *a posteriori* state probabilities $\gamma_{m,i}^{(\ell)}$ via the Forward-Backward algorithm [39] given the LP-HMM parameters and $X_m^{(\ell)}$, $m = 0, \ldots, M$.
4. Get the updated the estimate of $\alpha_1^{(\ell+1)}$ by applying the re-estimation formula (18)–(22).
5. Set $\ell = \ell + 1$ and go to 2 unless convergence is reached.
6. Derive the reverberation time T_{60} from $\alpha_1^{(\ell)}$ via (6).

reverberation time T_{60} via Eq. (6), is summarized in Table 2. Note that the Viterbi approximation [38, 39] may be postulated, taking into account only the most likely state path $s_{0,M}$ instead of considering all the possible state sequences via the Forward-Backward algorithm. In this case, the *a posteriori* state probabilities are constrained to be null except for one state at every time instant. This results in a computationally more efficient but suboptimal estimation algorithm.

3.3. Convergence Results

The proposed algorithm for the estimation of T_{60} has been derived in the EM framework. We can thus expect it to be convergent, if not for the approximations that had to be made. In this section, we do not mathematically prove the convergence of the estimation algorithm but rather report some results which demonstrate its convergence properties in practice.

First, we show by an example the convergence of the estimation algorithm for various values of the distortion parameter α_1. For every estimation trial, 256 observations are produced by applying the distortion model (see Section 2) to emissions drawn from the LP-HMM for anechoic speech (see Section 3.1). Actually, α_0 is set equal to 1 and five values of α_1 ranging from -0.45 to -0.85 are tested. The estimation algorithm is initialized with $\alpha_1^{(0)} = -0.95$. The iterative estimation procedure is stopped when two successive estimates do not differ by more than $\delta = 10^{-4}$ or if a maximum number $\ell_{\max} = 128$ of iterations is reached. Figure 5 shows the evolution of the estimate $\alpha_1^{(\ell)}$ as a function of the number ℓ of iterations. The estimation

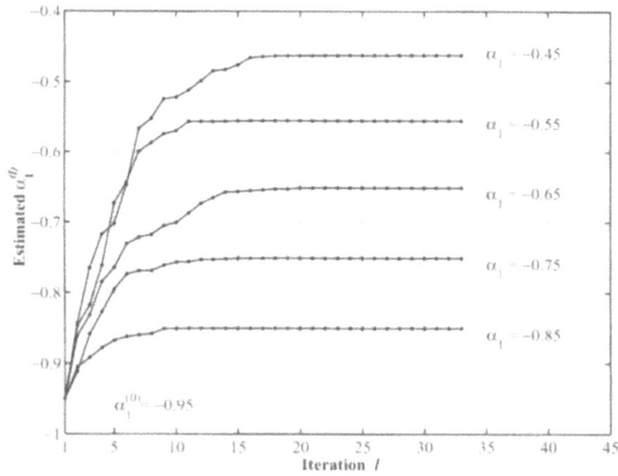

Figure 5. Convergence plot of the estimation algorithm for various values of the distortion parameter α_1. The length $M + 1$ of the observation sequence is set to 256.

algorithm clearly converges and an accurate estimate of the distortion parameter α_1 is obtained after a few iterations only.

Next, we run a series of Monte-Carlo experiments. Observations are again generated by applying the distortion model (α_1 varying from -0.45 to -0.85) to emissions of the anechoic speech LP-HMM ($M + 1$ varying from 8 to 512). The initialization and the stopping criterion are as described above. For every setup, 1024 Monte-Carlo trials are performed and the mean squared error (MSE) between the estimated distortion parameter $\alpha_1^{(\ell+1)}$ after convergence and the true one

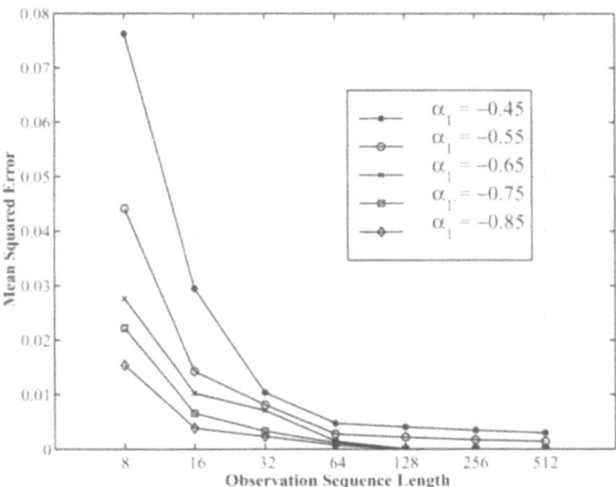

Figure 6. Mean squared error (MSE) for the distortion parameter α_1 as a function of the observation sequence length $M + 1$ and for various true values of the distortion parameter.

α_1 is computed. Figure 6 shows that the estimation algorithm is consistent, that is, the MSE value decreases when the number of observations increases. We clearly observe that the more observations the better the estimate, especially for high values of the distortion parameter α_1.

4. Speech Recognition Results

In this section, we report some results of our model selection approach for automatic speech recognition in reverberant environments. We first describe the experimental setup. Next, we give ASR results for both simulated reverberated speech and for real data recorded in reverberant enclosures.

4.1. Experimental Setup

The speech database used in this work comes from the noise-free part of the AURORA [41] corpus. It consists of connected digit sequences and it is divided into a training set of 8840 utterances and a test set of 1001 utterances, pronounced by 110 speakers and 104 other speakers, respectively. Recognition experiments are performed with a phoneme-based hybrid system relying on the Hidden Markov Model/Multilayer Perceptron (HMM/MLP) paradigm [43]. That is, a MLP is used for the acoustic modeling stage. The MLP is fed with acoustic features computed from 30 ms long/10 ms overlapping frames of speech signal sampled at 8 kHz. For every frame, it estimates the phoneme *a posteriori* probabilities. In the ASR experiments, three types of acoustic features are used: Mel-warped frequency cepstral coefficients (MFCC), MFCC with cepstral mean subtraction (CMS) [23] and logRASTA-PLP [24] coefficients. The last two front-ends are known to be robust to channel distortion. Speech decoding is done by Viterbi search, with neither pruning nor grammar constraints.

As described previously, we need to build a library of acoustic models (MLP's) trained on speech databases corresponding to various reverberant conditions. We first train a MLP on anechoic speech. Then, we train eight MLP's on artificially reverberated training sets. Training material is obtained with the reverberation process described in [3] for eight values of T_{60} varying uniformly from 200 ms to 1600 ms. All MLP's are obtained with the classical supervised back-propagation algorithm [43].

Once the library of acoustic models have been generated, we use them to recognize various reverberated test sets by selecting the most appropriate model. The selection is based on the estimation of the fullband reverberation time T_{60} by the method described in Section 3. Like all iterative methods, our T_{60} estimation algorithm requires an initial estimate $\alpha_1^{(0)}$ (see Section 3.2), and a stopping criterion. In practice, the reverberation time is lower than 2 seconds in most real reverberant enclosures. This upper bound is used to compute the initial estimate via Eq. (6), that is, $\alpha_1^{(0)} = -0.933$. The stopping criterion is the one described in Section 3.3. The LP-HMM modeling short-term log-energy sequences (see Section 3.1) is trained on a 64-utterance subset of the anechoic training set.

4.2. Results on Simulated Speech Data

Our model selection method is first tested on simulated reverberated speech. The test sets are obtained by convolving anechoic speech with room impulse responses computed by the Image Method [35, 36]. The main advantage of the room simulation approach is the flexibility and the control it offers on the reverberation time. Various test sets are generated within a 9 m long × 6 m wide × 4 m high simulated room. The microphone is located 2 m away from the speech source. The wall absorption coefficients are set such that T_{60} is ranging from 200 ms to 1600 ms.

First, we report cross-testing results (see Table 3). We see that the lowest word error rate (WER), i.e., the sum of the substitution (SUB), deletion (DEL) and insertion (INS) error rates, is always achieved by the acoustic model most closely matching the testing conditions (along main diagonal). Note that, as could have been expected, WER increases for the matching acoustic model when the reverberation becomes stronger.

Next, we apply our model selection approach by blind estimation of T_{60}. For all test sets, each reverberated speech utterance is processed prior to its recognition: the speech signal amplitude is normalized, the short-term log-energy sequence is computed for $T_w = 30$ ms and $F_r = 100$ Hz, T_{60} is estimated and the most closely matching MLP of the library is activated. Note that the blind T_{60} estimation algorithm becomes ill-conditioned for anechoic speech, but since real rooms are never anechoic, this is not a problem in practice.

Table 4 reports the performance of the selection method in terms of median absolute value errors (MAVE) for T_{60} estimation and confusion rates (CR), i.e., wrong acoustic model selection rates. The estimation algorithm performs satisfactorily and the most closely matching MLP is very often selected. The proposed model selection method approaches the recognition performance of the "Oracle" method (see Table 5) for which T_{60} is assumed to be known in advance and the best model is always selected. Clearly, the model selection method outperforms systems based on standard channel-robust acoustic features like MFCC-CMS

Table 3. Performances (WER (%)) of MFCC-based acoustic models trained on artificially reverberated speech for various simulated reverberant testing conditions.

Test set		Training set							
	Anechoic	$T_{60} = 200$ ms	400	600	800	1000	1200	1400	1600
Anechoic	**1.7**	2.9	7.6	11.9	15.9	19.8	20.6	22.7	23.8
$T_{60} = 200$ ms	7.0	**3.6**	4.5	6.4	9.8	12.5	13.7	15.1	16.1
300	7.8	**3.9**	4.4	6.4	9.8	12.3	13.9	15.0	15.7
400	18.7	9.6	**5.2**	5.7	8.5	12.2	12.8	14.7	15.3
500	20.1	11.2	**5.9**	5.9	8.7	12.2	12.8	14.7	15.4
600	29.7	20.2	11.3	**9.2**	10.0	12.6	13.7	15.6	16.4
700	33.2	24.7	14.9	**11.2**	11.3	13.6	14.4	16.1	17.1
800	41.0	33.7	22.1	17.3	**14.0**	15.9	16.6	18.2	19.1
1000	43.4	35.8	24.0	20.4	16.0	**17.0**	17.1	18.7	19.7
1200	49.3	43.1	32.0	27.9	20.9	20.7	**20.4**	21.6	22.1
1400	51.1	48.5	36.8	33.5	26.0	24.9	23.2	**24.5**	24.4
1600	52.9	50.1	37.3	36.6	28.1	26.7	25.3	25.1	**25.1**

Table 4. Median absolute value error (MAVE) (ms) and relative MAVE (%) for blind estimation of T_{60} and corresponding confusion rate (CR) (%) for model selection in the case of simulated reverberated speech.

Test set	MAVE (ms)	Relative MAVE (%)	CR (%)
200 ms	53.4	29.2	3.3
300 ms	37.8	12.6	1.2
400 ms	67.9	17.0	10.8
500 ms	64.1	12.8	9.0
600 ms	70.8	11.8	22.7
800 ms	88.7	11.1	10.7
1000 ms	52.4	5.2	20.6
1200 ms	95.0	7.9	47.8
1400 ms	103.8	7.4	51.1
1600 ms	130.3	8.1	31.8
Average	76.4	12.3	20.9

Table 5. Comparison of the performances (WER (SUB/DEL/INS) (%)) of the baseline system, two standard normalization techniques, the T_{60}-based model selection method and the "Oracle" method for simulated reverberated speech. Scores are averaged over all testing environments.

Method	WER (%)	SUB/DEL/INS (%)
MFCC-baseline	32.2	9.9/18.6/3.7
MFCC-CMS	37.5	8.7/27.4/1.4
logRASTA-PLP	36.1	11.0/24.1/1.0
T_{60}-based mod. sel.	13.3	4.7/5.9/2.7
"Oracle" mod. sel.	12.7	4.6/5.2/2.9

and logRASTA-PLP. These frame-based front-ends are not effective for handling reverberation because they have been designed to reduce the effect of short-time spectral coloration. However, the duration of acoustic impulse responses is typically longer than the temporal analysis window (e.g., 30 ms). Therefore, the reverberation effect is no longer multiplicative in the short-term spectral domain and cannot be considered as spectral coloration.

4.3. Results on Real Speech Data

Our model selection method has been also tested on speech recorded in four different reverberant enclosures, namely the "office", "meeting", "lavatory" and "cafeteria" enclosures. The names of the enclosures are self-explanatory. For every reverberant environment, the anechoic test set is played back with a loudspeaker

Table 6. Fullband reverberation times T_{60} (ms) measured by Schroeder's method in real reverberant environments.

Room	Description	T_{60} (ms)
Office	Medium size room with windows and concrete floor and walls	604.0
Meeting	Medium size room with glass windows and cement floor	873.3
Lavatory	Small room with tiled floor and walls	1307.1
Cafeteria	Large room with wooden floor and glass walls	1461.8

and simultaneously recorded with a distant omnidirectional microphone. The distance between the loudspeaker and the microphone is set to 2 meters. Room impulse responses are also measured using a correlation method based on an optimal time-stretched pulse (TSP) [44]. No attempt is made to compensate for the transfer functions of the loudspeaker and the microphone since both are studio-grade devices. The fullband reverberation times are estimated from Schroeder's decay curves [33] (i.e., using the time-reverse integrals of the squared room impulse responses) and are reported in Table 6.

First, we report cross-testing results (see Table 7). The same conclusions as for simulated reverberated speech can be drawn: the best results are obtained when the acoustic model most closely matching the reverberation time is activated. Unfortunately, we observe that the recognition performance are not as good as for simulated reverberated speech when the training conditions match exactly the testing conditions. e.g., for the "office" environment. Moreover, the performance degrade more severely if the conditions do not match. These differences can be explained by noting that the reverberation time T_{60} is not independent of the frequency while our reverberation process for training databases assumes so. Indeed, most real materials are more absorptive at high frequency and the absorption property of the air increases with frequency. Consequently, the reverberation time decreases with frequency in real acoustic environments leading to a mismatch with our proposed model. Nevertheless, our model selection method has been applied to the four real reverberated test sets. The performance of the estimation algorithm are reported in Table 8. Although the estimation algorithm does not perform as well as for simulated reverberated speech, the model selection method still yields significant improvements over the baseline ASR system (see Table 9).

Table 7. Performances (WER (%)) of MFCC-based acoustic models trained on artificially reverberated speech for various real reverberant testing conditions.

Test set	Anechoic	$T_{60} = 200$ ms	400	600	800	1000	1200	1400	1600
					Training set				
Anechoic	**1.7**	2.9	7.6	11.9	15.9	19.8	20.6	22.7	23.8
Office	30.2	24.8	13.8	**11.2**	15.6	19.0	21.2	22.7	27.2
Meeting	47.7	44.8	29.2	23.0	**20.5**	23.7	28.0	28.0	30.4
Lavatory	57.9	59.2	50.5	47.9	43.7	39.6	36.9	**33.9**	37.9
Cafeteria	57.8	56.4	47.9	42.7	38.4	35.6	36.1	**32.9**	37.7

Table 8. Median absolute value error (MAVE) (ms) and relative MAVE (%) for blind estimation of T_{60} and corresponding confusion rate (CR) (%) for model selection in the case of real reverberated speech.

Test set	MAVE (ms)	Relative MAVE (%)	CR (%)
Office	143.8	23.8	34.0
Meeting	120.8	13.8	21.8
Lavatory	126.1	9.7	19.9
Cafeteria	137.4	9.4	22.8
Average	132.0	14.2	24.6

Table 9. Comparison of the performances WER (SUB/DEL/INS) (%) of the baseline system, two standard normalization techniques, the T_{60}-based model selection method and the "Oracle" method for real reverberated speech. Scores are averaged over all testing environments.

Method	WER (%)	SUB/DEL/INS (%)
MFCC-baseline	48.4	22.3/21.8/4.2
MFCC-CMS	53.3	14.6/36.9/1.8
logRASTA-PLP	51.9	19.9/27.2/4.8
T_{60}-based mod. sel.	28.2	14.7/8.8/4.7
"Oracle" mod. sel.	24.7	12.8/7.5/4.4

5. Summary and Concluding Remarks

In this communication, we have proposed a model selection method for automatic speech recognition in reverberant rooms. Our method consists in selecting the best acoustic model out of a library of models trained separately on artificially reverberated speech databases matching various reverberation times. The selection is based on the estimation of the fullband reverberation time. The estimation algorithm computes a Maximum Likelihood estimate of the reverberation time using only the short-term log-energy sequence of a reverberated speech utterance recorded in the reverberant operating environment.

Experimental recognition tests on connected digit sequences have shown that the estimation algorithm performs well and the model selection approach improves significantly recognition accuracy for both simulated and real reverberated data. However, the performance improvement is not as high as expected on real data. This can be explained by the fact that the current modeling and estimation methods assume that the acoustic field is diffuse and that the reverberation time is independent of frequency. Further improvements can be expected by relaxing these severe assumptions. For example, we might consider estimating the reverberation time in frequency subbands. The training procedure should be modified accordingly.

Another shortcoming of the current model selection approach to hands-free speech recognition is that it deals only with reverberation: the speech signal is supposed to be noiseless. The proposed method should be extended to noisy environments to constitute a complete solution. This will be the subject of future work.

Appendix A: Linear Distortion Model

Consider the distortion model of Fig. 2 and the LP-HMM of Fig. 4. Define $G(\cdot)$ as the transformation related the observation sequence Y_m to the LP-HMM emission sequence E_m given a state sequence s_m. The transformation $G(\cdot)$ is given by the following equations:

$$E_m = \sum_{p=0}^{1} \beta_p(s_m) X_{m-p}, \quad X_m = 10 \log_{10}(W_m),$$

$$W_m = \sum_{k=0}^{1} \alpha_k Z_{m-k}, \quad Z_m = 10^{Y_m/10}. \tag{23}$$

Clearly, the transformation is non-linear because of the logarithm and power functions. In Section 3, we describe an iterative method for the ML estimation of the distortion parameters (α_0, α_1) based on the EM algorithm. In order to derive close-form re-estimation formulae, the transformation of Eq. (23) has to be linearized with respect to the distortion parameters. Let us assume that the successive estimates of the distortion parameters are varying slowly enough such that the sequence $W_m^{(\ell)} = \alpha_0^{(\ell)} Z_m + \alpha_1^{(\ell)} Z_{m-1}$ estimated at ℓ-th step is close to the next estimate $W_m^{(\ell+1)} = \alpha_0^{(\ell+1)} Z_m + \alpha_1^{(\ell+1)} Z_{m-1}$. Using a Taylor series expansion limited to the first order, we get the following linear approximation:

$$10 \log_{10}\left(W_m^{(\ell+1)}\right) \simeq 10 \log_{10}\left(W_m^{(\ell)}\right) + \left(W_m^{(\ell+1)} - W_m^{(\ell)}\right)/\xi W_m^{(\ell)}, \quad \xi = \log(10)/10. \tag{24}$$

Equations (23) becomes

$$E_m^{(\ell+1)} \simeq \sum_{p=0}^{1} \beta_p(s_m) \left[\sum_{k=0}^{1} A_{k,m-p}^{(\ell+1)} Z_{m-p-k} + B_{m-p}^{(\ell+1)} \right] \tag{25}$$

with

$$A_{k,m}^{(\ell+1)} \triangleq \alpha_k^{(\ell+1)}/\xi W_m^{(\ell)} \tag{26}$$

and

$$B_m^{(\ell+1)} \triangleq X_m^{(\ell)} - 1/\xi. \tag{27}$$

If we neglect the boundary effects, that is, if we assume $Z_m = 0$ for $m < 0$ and $m > M$, Eq. (25) can be written in a matrix form for a finite-length sequence of observations $Z_{0,M} = \{Z_0, \ldots, Z_M\}$ and the corresponding finite-length sequence $E_{0,M}$ of emissions:

$$E_{0,M}^{(\ell+1)} = G\left(Y_{0,M} \mid s_{0,M}, \alpha^{(\ell+1)}\right)$$
$$\simeq \mathbf{B}^{(\ell+1)}\left[\mathbf{A}^{(\ell+1)} Z_{0,M} + \mathbf{b}^{(\ell+1)}\right], \tag{28}$$

where the $(M+1) \times (M+1)$ matrices $\mathbf{B}$ and $\mathbf{A}$, and the $(M+1) \times 1$ vector $\mathbf{b}$ are defined by,

$$\mathbf{B}^{(\ell+1)} \triangleq \begin{bmatrix} \beta_0(s_0) & 0 & \cdots & & 0 \\ \beta_1(s_1) & \beta_0(s_1) & & & \vdots \\ 0 & \ddots & \ddots & & \\ \vdots & & & & 0 \\ 0 & \cdots & & \beta_1(s_m) & \beta_0(s_m) \end{bmatrix}, \tag{29}$$

$$\mathbf{A}^{(\ell+1)} \triangleq \begin{bmatrix} A_{0,0}^{(\ell+1)} & 0 & \cdots & & 0 \\ A_{1,1}^{(\ell+1)} & A_{0,1}^{(\ell+1)} & & & \vdots \\ 0 & \ddots & \ddots & & \\ \vdots & & & & 0 \\ 0 & \cdots & & A_{1,M}^{(\ell+1)} & A_{0,M}^{(\ell+1)} \end{bmatrix}, \tag{30}$$

and

$$\mathbf{b}^{(\ell+1)} \triangleq \begin{bmatrix} B_0^{(\ell+1)} \\ B_1^{(\ell+1)} \\ \vdots \\ B_M^{(\ell+1)} \end{bmatrix}, \tag{31}$$

respectively.

The approximation of the transformation $G(Y_{0,M} \mid s_{0,M}, \alpha)$ is linear with respect to the distortion parameters $\alpha = (\alpha_0, \alpha_1)$. Using Eqs. (23) and (28), its Jacobian matrix $\mathbf{J}$ can be computed as

$$\mathbf{J} \triangleq \frac{\partial G(Y_{0,M} \mid s_{0,M}, \alpha^{(\ell+1)})}{\partial Y_{0,M}}$$
$$= \frac{\partial G(\cdot)}{\partial Z_{0,M}} \cdot \frac{\partial Z_{0,M}}{\partial Y_{0,M}} = \mathbf{B}^{(\ell+1)} \mathbf{A}^{(\ell+1)} \mathbf{D}, \tag{32}$$

with $\mathbf{D}$ denoting a diagonal matrix whose elements are equal to $\xi Y_{0,M}$. It is straightforward to verify that the determinant of the Jacobian matrix $\mathbf{J}$ is given by

$$|\mathbf{J}| = \left|\mathbf{B}^{(\ell+1)}\right|\left|\mathbf{A}^{(\ell+1)}\right||\mathbf{D}|$$
$$= \prod_{m=0}^{M} \beta_0(s_m) A_{0,m}^{(\ell+1)} \xi Y_m. \tag{33}$$

References

1. R. Siemund, H. Höge, S. Kunzmann, and K. Marasek, "SPEECON—Speech Data for Consumer Devices," in *Proc. of International Conference on Language Resources and Evaluation (LREC)*, Athens, Greece, 2000, vol. 2, pp. 883–886.
2. S. Nakamura and K. Shikano, "Room Acoustics and Reverberation: Impact on Hands-Free Recognition," in *Proc. of European Conference on Speech Communication and Technology (EUROSPEECH)*, Rhodes, Greece, 1997, vol. 5, pp. 2419–2422.
3. L. Couvreur, C. Couvreur, and C. Ris, "A Corpus-Based Approach for Robust ASR in Reverberant Environments," in *Proc. of International Conference on Spoken Language Processing (ICSLP)*, Beijing, China, 2000, vol. 1, pp. 397–400.

4. Y. Pan and A. Waibel, "The Effects of Room Acoustics on MFCC Speech Parameter," in *Proc. of International Conference on Spoken Language Processing (ICSLP)*, Beijing, China, 2000.

5. C. Avendano and H. Hermansky, "Study on the Dereverberation of Speech Based on Temporal Envelope Filtering," in *Proc. of International Conference on Spoken Language Processing (ICSLP)*, Philadelphia, USA, 1996, vol. 2, pp. 889–892.

6. S. Subramaniam, A.P. Petropulu, and C. Wendt, "Cepstrum-Based Deconvolution for Speech Dereverberation," *IEEE Trans. on Speech and Audio Processing*, vol. 4, no. 5, 1996, pp. 392–396.

7. D. Cole, M. Moody, and S. Sridharan, "Position-Independent Enhancement of Reverberant Speech," *Journal of Audio Engineering Society*, vol. 45, no. 3, 1997, pp. 142–147.

8. H. Nomura, S. Hirobayashi, T. Koike, and M. Tohyama, "Dereverberation of Speech by Power Envelope Inverse Filtering," in *Proc. of IEEE Workshop on Digital Signal Processing*, Bryce Canyon, USA, 1998, vol. 1, pp. 229–232.

9. B. Yegnanarayana, P.M. Satyanarayanan, C. Avendano, and H. Hermansky, "Enhancement of Reverberant Speech Using LP Residual," in *Proc. of International Conference on Acoustics, Speech and Signal Processing (ICASSP)*, Seattle, USA, 1998, vol. 1, pp. 405–408.

10. Q.-G. Liu, B. Champagne, and P. Kabal, "A Microphone Array Processing Technique for Speech Enhancement in Reverberant Space," *Speech Communication*, vol. 18, no. 4, 1996, pp. 317–334.

11. C. Marro, Y. Mahieux, and K.U. Simmer, "Analysis of Noise Reduction and Dereverberation Techniques Based on Microphone Arrays with Postfiltering," *IEEE Trans. on Speech and Audio Processing*, vol. 6, no. 3, 1998, pp. 240–259.

12. F. Asano, S. Hayamizu, T. Yamada, and S. Nakamura, "Speech Enhancement Based on the Subspace Method," *IEEE Trans. on Speech and Audio Processing*, vol. 8, no. 5, 2000, pp. 497–507.

13. A. Koutras, E. Dermatas, and G. Kokkinakis, "Improving Simultaneous Speech Recognition in Real Room Environments Using Overdetermined Blind Source Separation," in *Proc. of European Conference on Speech Communication and Technology (EUROSPEECH)*, Aalborg, Denmark, 2001, vol. 2, pp. 1009–1013.

14. R. Mukai, S. Araki, and S. Makino, "Separation and Dereverberation Performance of Frequency Domain Blind Source Separation for Speech in a Reverberant Environment," in *Proc. of European Conference on Speech Communication and Technology (EUROSPEECH)*, Aalborg, Denmark, 2001, vol. 4, pp. 2599–2602.

15. B.D. Radlovic̀, R.C. Williamson, and R.A. Kennedy, "Equalization in an Acoustic Reverberant Environment: Robustness Results," *IEEE Trans. on Speech and Audio Processing*, vol. 8, no. 3, 2000, pp. 311–319.

16. S.T. Neely and J.B. Allen, "Invertibility of a Room Impulse Response," *Journal of the Acoustical Society of America*, vol. 66, no. 1, 1979, pp. 165–169.

17. M. Matassoni, M. Omologo, and D. Giuliani, "Hands-Free Speech Recognition Using a Filtered Clean Corpus and Incremental HMM Adaptation," in *Proc. of International Conference on Acoustics, Speech and Signal Processing (ICASSP)*, Istanbul, Turkey, 2000, vol. 3, pp. 1407–1410.

18. T. Takiguchi, S. Nakamura, and K. Shikano, "HMM-Separation-Based Speech Recognition for a Distant Moving Speaker," *IEEE Trans. on Speech and Audio Processing*, vol. 9, no. 2, 2001, pp. 127–140.

19. L. Couvreur, S. Dupont, C. Ris, J.-M. Boite, and C. Couvreur, "Fast Adaptation for Robust Speech Recognition in Reverberant Environments," in *Proc. of ISCA Workshop on Adaptation Methods For Automatic Speech Recognition*, Sophia Antipolis, France, 2001, pp. 85–88.

20. L. Rigazio, D. Kryze, P. Nguyen, and J.-C. Junqua, "Joint Environment and Speaker Adaptation," in *Proc. of ISCA Workshop on Adaptation Methods For Automatic Speech Recognition*, Sophia Antipolis, France, 2001, pp. 93–96.

21. K. Yamamoto, S. Nakagawa, and H. Matsumoto, "Evaluation of PMC for Segmental Unit Input HMM in Various Environments," in *Proc. of International Workshop on Hands-free Speech Communication*, Kyoto, Japan, 2001, pp. 183–186.

22. Y. Zhao, "Statistical Estimation for Hands-Free Speech Recognition," in *Proc. of International Workshop on Hands-free Speech Communication*, Kyoto, Japan, 2001, pp. 183–186.

23. S. Furui, "Cepstral Analysis Technique for Automatic Speaker Verification," *IEEE Trans. on Acoustics, Speech and Signal Processing*, vol. 29, no. 2, 1981, pp. 254–272.

24. H. Hermansky and N. Morgan, "RASTA Processing of Speech," *IEEE Trans. on Speech and Audio Processing*, vol. 2, no. 4, 1994, pp. 578–589.

25. C. Avendano, S. Van Vuuren, and H. Hermansky, "Data Based Filter Design for RASTA-like Channel Normalization in ASR," in *Proc. of International Conference on Spoken Language Processing (ICSLP)*, Philadelphia, USA, 1996, vol. 4, pp. 2087–2090.

26. B. Kingsbury and N. Morgan, "Recognizing Reverberant Speech with RASTA-PLP," in *Proc. of International Conference on Acoustics, Speech and Signal Processing (ICASSP)*, Munich, Germany, 1997, vol. 2, pp. 1259–1262.

27. B. Kingsbury, N. Morgan, and S. Greenberg, "Improving ASR Performance For Reverberant Speech," in *Proc. of ESCA Workshop on Robust Speech Recognition for Unknown Communication Channels*, Pont-à-Mousson, France, 1997, pp. 87–90.

28. M.L. Shire and B.Y. Chen, "Data-Driven RASTA Filters in Reverberation," in *Proc. of International Conference on Acoustics, Speech and Signal Processing (ICASSP)*, Istanbul, Turkey, 2000, vol. 3, pp. 1627–1630.

29. M.L. Shire and B.Y. Chen, "On Data-derived Temporal Processing in Speech Feature Extraction," in *Proc. of International Conference on Spoken Language Processing (ICSLP)*, Beijing, China, 2000.

30. D. Giuliani, M. Matassoni, M. Omologo, and P. Svaizer, "Training of HMM with Filtered Speech Material for Hands-free Recognition," in *Proc. of International Conference on Acoustics, Speech and Signal Processing (ICASSP)*, Phoenix, USA, 1999, vol. 1, pp. 449–452.

31. V. Stahl, A. Fischer, and R. Bippus, "Acoustic Synthesis of Training Data for Speech Recognition in Living Room Environments," in *Proc. of International Conference on Acoustics, Speech and Signal Processing (ICASSP)*, Salt Lake City, USA, 2001, vol. 1, pp. 21–24.

32. M. Omura, M. Yada, H. Saruwatari, S. Kajita, K. Takeda, and F. Itakura, "Compensating of Room Acoustic Transfer Functions Affected by Change of Room Temperature," in *Proc. of International Conference on Acoustics, Speech and Signal Processing (ICASSP)*, Phoenix, USA, 1999, vol. 2, pp. 941–944.

33. H. Kuttruff, *Room Acoustics*, 4th ed. Elsevier, 2000.

34. A. Sankar and C.-H. Lee, "A Maximum-Likelihood Approach to Stochastic Matching for Robust Speech Recognition," *IEEE Trans. on Speech and Audio Processing*, vol. 4, no. 3, 1996, pp. 190–202.

35. J.B. Allen and D.A. Berkley, "Image Method for Efficiently Simulating Small-Room Acoustics," *Journal of the Acoustical Society of America*, vol. 65, no. 4, 1979, pp. 943–950.

36. P.M. Peterson, "Simulating the Response of Multiple Microphones to a Single Acoustic Source in a Reverberant Room," *Journal of the Acoustical Society of America*, vol. 80, no. 5, 1986, pp. 1527–1529.

37. J. Moorer, "About this Reverberation Business," *Computer Music Journal*, vol. 3, no. 2, 1979, pp. 13–28.

38. C.J. Wellekens, "Explicit Time Correlation in Hidden Markov Models for Speech Recognition," in *Proc. of International Conference on Acoustics, Speech and Signal Processing (ICASSP)*, Dallas, USA, 1987, vol. 1, pp. 384–387.

39. P. Kenny, M. Lennig, and P. Mermelstein, "A Linear Predictive HMM for Vector-Valued Observations with Applications to Speech Recognition," *IEEE Trans. on Acoustics, Speech and Signal Processing*, vol. 38, no. 2, 1990, pp. 220–225.

40. A.P. Dempster, N.M. Laird, and D.B. Rubin, "Maximum Likelihood from Incomplete Data via the EM Algorithm," *Journal of the Royal Statistical Society*, ser. B, vol. 39, 1997, pp. 1–38.

41. AURORA database, http://www.elda.fr/aurora2.html.

42. A. Papoulis, *Probability, Random Variables, and Stochastic Processes*, 3rd ed. McGraw-Hill, 1991.

43. H. Bourlard and N. Morgan, *Connectionist Speech Recognition—A Hybrid Approach*, Kluwer Academic Publishers, 1994.

44. Y. Suzuki, F. Asano, H.-Y. Kim, and T. Sone, "An Optimum Computer-Generated Pulse Signal Suitable for the Measurement of Very Long Impulse Responses," *Journal of the Acoustical Society America*, vol. 97, no. 2, 1995, pp. 1119–1123.

Laurent Couvreur was born in Nivelles, Belgium, in 1974. He received the Electrical Engineering (Ir) degree with highest honors from the Faculté Polytechnique de Mons, Belgium, in 1997. Since 1999, he has been a Ph.D. student in the Signal Processing Laboratory at Faculté Polytechnique de Mons. From January 2000 to August 2000, he was a Visiting Scholar in the Coordinated Science Laboratory at the University of Illinois at Urbana-Champaing. He is now also working in Multitel, as a research engineer in the automatic speech recognition group.

His current research interests include statistical signal processing, audio and speech processing, and more especially, automatic speech recognition in adverse conditions.
lcouv@tcts.fpms.ac.be

Christophe Couvreur was born in Nivelles, Belgium, in 1968. He received the Engineering (Ir) degree summa cum laude from the Faculté Polytechnique de Mons, Belgium, in 1991 and the M.Sc. degree from the University of Illinois at Urbana-Champaign, USA, in 1995, both in Electrical Engineering. He also received the Diplôme dÉtudes Approfondies (DEA) degree in Mathematics from the Université Catholique de Louvain, Belgium, in 1996 and the Doctorat en Sciences Appliquées (Ph.D.) degree from the Faculté Polytechnique de Mons in 1997.

Since 2001, Christophe Couvreur has been with the Speech and Language Technology Division of ScanSoft, Inc., where he is a Principal Research Engineer working on speech recognition systems. Prior to joining ScanSoft, Christophe Couvreur held research positions with the Belgian National Fund for Scientific Research (FNRS), with the University of Illinois at Urbana-Champaign, and with Lernout & Hauspie Speech Products. He also served as an officer in the Belgian Air Force.

His current research interests include speech recognition, speech and audio processing, pattern recognition, statistical signal and array processing, and software engineering.
Christophe.Couvreur@scansoft.com

FSC
www.fsc.org
MIX
Papier aus verantwortungsvollen Quellen
Paper from responsible sources
FSC® C105338